SYSTEMATICS AND THE FOSSIL RECORD

Systematics and the fossil record:
documenting evolutionary patterns

ANDREW B. SMITH

The Natural History Museum
Cromwell Road, London

OXFORD

BLACKWELL SCIENTIFIC PUBLICATIONS

LONDON EDINBURGH BOSTON

MELBOURNE PARIS BERLIN VIENNA

© 1994 by
Blackwell Scientific Publications
Editorial Offices:
Osney Mead, Oxford OX2 0EL
25 John Street, London WC1N 2BL
23 Ainslie Place, Edinburgh EH3 6AJ
238 Main Street, Cambridge
    Massachusetts 02142, USA
54 University Street, Carlton
    Victoria 3053, Australia

Other Editorial Offices:
Librairie Arnette SA
1, rue de Lille
75007 Paris
France

Blackwell Wissenschafts-Verlag GmbH
Düsseldorfer Str. 38
D-10707 Berlin
Germany

Blackwell MZV
Feldgasse 13
A-1238 Wien
Austria

First published 1994

Set by Excel Typesetters, Hong Kong
Printed and bound in Great Britain
at The Alden Press, Oxford

DISTRIBUTORS

Marston Book Services Ltd
PO Box 87
Oxford OX2 0DT
(*Orders*: Tel: 0865 791155
        Fax: 0865 791927
        Telex: 837515)

USA
Blackwell Scientific Publications, Inc.
238 Main Street
Cambridge, MA 02142
(*Orders*: Tel: 800 759-6102
            617 876-7000)

Canada
Oxford University Press
70 Wynford Drive
Don Mills
Ontario M3C 1J9
(*Orders*: Tel: 416 441-2941)

Australia
Blackwell Scientific Publications
    Pty Ltd
54 University Street
Carlton, Victoria 3053
(*Orders*: Tel: 03 347-5552)

A catalogue record for this title is
available from the British Library

ISBN 0-632-03642-7

Library of Congress
Cataloging-in-Publication Data

Smith, Andrew B.
    Systematics and the fossil record:
documenting evolutionary patterns /
    Andrew B. Smith.
        p.   cm.
    Includes bibliographical references
and index.
    ISBN 0-632-03642-7
    1. Evolutionary paleobiology.
    2. Cladistic analysis. I. Title.
QE721.2.E85S65   1994
560 – dc20

Contents

Preface

It is the study of fossils that provides the most direct evidence we have for how evolution has proceeded over geological time. Yet evolutionary history cannot simply be read from the rocks. Individual fossil finds must first be grouped and rationalized into a coherent framework to make sense of the record. This is done through the application of systematics. It is through systematics that we recognize the basic biological units from which all evolutionary patterns are deduced, and it is through systematics that we construct phylogenetic hypotheses and higher taxa. We therefore need to use the best available systematic methods if we are to improve the accuracy of evolutionary patterns, and so sharpen our ideas of evolutionary processes.

During the past decade the theory and method of cladistics has developed rapidly, and it is now widely acknowledged to be the best tool available for establishing phylogenetic relationships. Furthermore it can compensate for sampling and preservation biases more fully than any other method. Yet investigation of evolutionary patterns in the fossil record has continued largely oblivious to changes in systematic practices. This book therefore sets out to explain why the study of evolutionary patterns and processes must begin from a firm taxonomic foundation. Our best estimate of the branching geometry of the tree of life comes from combining hypotheses of phylogenetic relationship derived from cladistic character analysis with biostratigraphical data.

In the following chapters I explain the theory and methodology behind this approach emphasizing the positive benefits that it brings. In doing so I have had to tread an honest path between the stringent logic of cladistic theory and the practical aspects of working with anatomically incomplete fossils and a spatiotemporally incomplete stratigraphical evidence.

I owe a great deal of thanks to those that have helped by offering suggestions and comments on an earlier draft of this book. Derek Briggs, Joel Cracraft, Richard Fortey, Colin Patterson and Paul Taylor all devoted time and effort to this task and their positive criticism has helped me to sharpen and clarify my arguments. For this I am extremely grateful. Needless to say, though, the views in this book reflect my own perspective alone and I am responsible for all errors that may have crept in.

A.B. Smith

1 Introduction

One of the most intriguing aspects of the fossil record is the way in which biodiversity has changed through geological time. Although species diversity overall has increased through the Phanerozoic (Sepkoski *et al.*, 1981; Signor, 1990), this has been achieved through a bewilderingly complex pattern of shifting dominance as the fortunes of myriads of individual clades have waxed and waned.

The underlying causes of this variation are clearly complex and must encompass everything from small-scale stochastic processes acting locally on individual species, through to major extrinsic factors acting on a global scale and thus affecting many clades simultaneously. Major extrinsic factors, such as sea-level change, climatic fluctuations, plate tectonics, and bolide impacts, have undoubtedly played some part in determining the history of life on this planet, and one of the fascinating tasks facing palaeontologists is to demonstrate the existence of general patterns of evolution. In order to do this we need accurate and reliable methods for reconstructing evolutionary patterns, free from artefact and bias. The two most widely used methods derive their primary data either from evolutionary trees or from taxonomic structure.

Methods for reconstructing evolutionary patterns

Taxonomy and the fossil record provide the two dimensions essential for reconstructing evolutionary patterns. It is through taxonomy that we identify species and establish how these are related, grouping them into appropriate higher taxa. And it is through the fossil record that we establish the times of appearance and disappearance of these species and higher taxa. If evolution was entirely the product of small-scale stochastic processes then the originations and extinctions of species and taxa should be randomly distributed through time. If, on the other hand, larger-scale extrinsic factors have been dominant, patterns of origination and/or extinction should be congruent across a wide spectrum of different clades.

However, evolutionary patterns cannot simply be read directly from the fossil record. The fossil record provides only a very imperfect record of past life, and sampling biases can grossly distort or mask the original signal. Somehow we need to overcome these sampling and preservational biases if we are to achieve a clearer understanding of evolutionary patterns.

There are two approaches that can help overcome biases in the fossil

1

record and which are currently used to discover evolutionary patterns: (i) the taxic approach, which derives its patterns from the structure of a classificatory scheme; and (ii) the phylogenetic approach, which identifies patterns from tree geometry.

The taxic approach

The study of evolutionary patterns really began in earnest with the development of a more quantitative approach to the fossil record. Earlier workers (e.g. Simpson, 1944, 1953; Cloud, 1948; Newell, 1952, 1967; Valentine, 1969) pioneered this approach using taxonomic data to investigate broad patterns of changing diversity through time. Subsequently it was firmly set forth as a major research programme under the title 'nomothetic palaeontology' (Raup *et al.*, 1973). Nomothetic palaeontology can be defined as the search for law-like relationships in evolution that can be derived from the distributional patterns of fossils in the geological record. The main tool that has been used to advance nomothetic palaeontology has been the analysis of taxonomic ranges, and this approach has dominated palaeobiological studies over the past decade. Levinton (1988) and others have dubbed this the 'taxic approach'.

The taxic approach starts from a formal classification. Some level in the taxonomic hierarchy (family, order, etc.) is selected and the earliest and latest records of any species classified within each higher taxon are used to define its total geological range. By compiling large databases of such range information, patterns of origination, standing diversity and extinction can be searched for and their significance tested using standard statistical techniques. The relationships of these taxa to one another are irrelevant because it is the co-occurrence of taxonomic appearances and disappearances that is important. Furthermore the taxonomy does not have to be perfectly accurate since the method will accommodate a certain amount of noisy data. Noise will tend only to degrade any signal that is present – or so it is hoped.

The taxic approach has several advantages. It uses large numbers of taxa and can therefore identify general trends from 'noisy' data where small samples might give misleading patterns. With large databases statistical tests can be applied, allowing the validity and robustness of patterns to be established. Finally, it is easy to apply, requiring minimal phylogenetic information: the relationships of taxa need not be known, and therefore evolutionary trees do not have to be fully worked out prior to the application of this method.

The taxic approach has been used to great effect in a wide range of studies aimed at discovering general patterns of evolution, e.g. (i) to identify major periods of mass extinction and determine whether they are materially different from 'background' extinction (Raup & Sepkoski, 1982, 1984; Sepkoski, 1984, 1986; Benton, 1985, 1986; Fox,

1987; Raup & Boyajian, 1988); (ii) to document patterns of global diversity through time and also how standing diversity within taxa of a given rank has changed (Sepkoski, 1978, 1984; Sepkoski *et al.*, 1981; Stanley, *et al.*, 1981; Flessa & Jablonski, 1985; Kitchell & Carr, 1985); (iii) to examine origination patterns (Flessa & Levinton, 1975; Gilinsky & Bambach, 1987); (iv) to document patterns of taxonomic duration (Levinton & Farris, 1987; Foote, 1988; Gilinsky, 1988; Gilinsky & Good, 1991); (v) to explore whether there are differences in evolutionary patterns between taxa of high and low rank (Valentine, 1969; Holman, 1989); (vi) to examine onshore–offshore patterns of origination (Jablonski *et al.*, 1983; Jablonski & Bottjer, 1990a,b, 1991); and (vii) to examine the effect of geographic distribution on survival (Jablonski, 1986a, 1987; Westrop & Ludvigsen, 1987; Westrop, 1991).

However, the method has to make two important assumptions about the taxa it uses. Firstly it assumes that the taxa used are real, not some arbitrary convention of taxonomists, and that their appearances and disappearances represent real biological events; secondly, it assumes that taxa of equal rank are approximately equivalent entities across diverse clades. Ultimately, therefore, the reliability of the approach stands or falls on the validity of the taxa used as its primary data. The weakness of the taxic approach lies in the failure of pre-cladistic systematists to generate a consistent taxonomy. Traditional databases are riddled with nonmonophyletic groupings that arise solely because of *ad hoc* taxonomic practice and have no claim to biological reality, and taxonomic rank has always been assigned arbitrarily.

Evolutionary pattern analysis is now, therefore, entering a new phase as the palaeontological database starts to undergo a major transformation.

Cladistics and the phylogenetic approach

With the development and widespread acceptance of cladistic methodology came the realization that many existing taxonomies were woefully inadequate. The cladistic revolution is now well under way and the taxonomic database is undergoing a major overhaul, with the eradication of non-monophyletic taxa and the recognition of taxonomic rank as a convention indicative only of level of inclusiveness.

The construction of cladistic hypotheses of relationship, based solely on the recognition of shared derived characters, has become widely accepted as the most powerful tool for investigating historical patterns. This is because it is the only method that will reconstruct relationships in a way that consistently reflects the phylogenetic history of a group. In recent years cladistic methods themselves have become more rigorous through the development of computer programs such as PAUP (Swofford, 1985, 1993) and HENNIG86 (Farris, 1988).

Cladistics has become transformed, to a large extent, into a school of numerical taxonomy.

The role of cladograms in historical biogeography is also now well established (Nelson & Platnick, 1981; Humphries & Parenti, 1986; Bremer, 1992) and is discussed further in Chapter 7. Biogeographic patterns that have arisen from major extrinsic events are distinguished from patterns of dispersal unique to individual lineages by searching for congruence of cladogram structure among diverse groups. Where the same biogeographic patterns recur in different taxonomic groups those patterns are likely to have been generated by common extrinsic factors. Exactly the same approach can now be applied to the discovery of general patterns of evolution through comparison of phylogenetic tree topology across different clades.

Palaeontological data and evolutionary trees

Partly in response to the excesses of palaeontological ancestor-hunting, which had previously dominated evolutionary methods, cladists initially dismissed any role for palaeontological data in phylogenetic reconstruction. Phylogenetic reconstruction was seen as a comparative method using morphological, biochemical, or genetic data. Since fossils are almost always less well known than extant organisms and the fossil record is notoriously incomplete, palaeontological data were thought to have no significant role in reconstructing phylogenies for extant taxa (Hennig, 1966; Nelson & Platnick, 1981; Patterson, 1982). However, there is now a growing realization that fossil data can be important for the accurate determination of phylogenetic relationships and character evolution (Donoghue *et al.*, 1989; Novacek, 1992a,b; Wilson, 1992). Our best estimates of phylogenetic relationships come from considering total evidence, and to omit fossil taxa is to ignore data that may in some cases be crucial. This theme is explored further in Chapter 3.

Cladistic methods have received a lot of attention, but there has been much less consideration given to the transformation of cladograms into phylogenetic trees. Phylogenetic trees are cladograms calibrated against the fossil record. They combine the results of morphological analysis with biostratigraphic data, using the fossil record to date the earliest appearances of taxa with derived characters and thus establish the time by which each branch of the cladogram must have come into existence. If the fossil record is good, these estimates should be reasonably accurate.

In the construction of phylogenetic trees morphological and stratigraphic data are kept clearly separate. Firstly phylogenetic relationships are established on the basis of comparative anatomy alone. Later, biostratigraphic data can be added to calibrate the cladogram, transforming it into a phylogenetic tree. The strength of this approach is that sampling bias can be compensated to a much greater extent than is possible by any other method. Integrating biostratigraphic evidence with that derived from comparative anatomy generates our best estimate of both the branching pattern and the branch lengths of the tree of life (see Chapter 6).

Phylogenetic trees are essential if we are to improve our understanding of evolutionary patterns and processes. They can be used to investigate a wide range of evolutionary questions. Chapter 7 outlines the way in which such trees can be used to document patterns of diversity, origination and extinction, rates of evolution, and biogeographic history in the fossil record.

The statistical approach, which is the strength of the taxic method, is still required to identify those patterns that have arisen because of some common extrinsic factor, but this can now be based on identifying congruent patterns of tree topology, not on pre-cladistic taxonomy of dubious validity. Evolutionary patterns that are the product of major extrinsic factors can be differentiated from small-scale stochastic processes by looking for congruence among different taxa. Simultaneous termination of branches across many clades, for example, if greater than that expected by chance, implies that there was a common extrinsic factor involved.

Trees are only as good as the cladograms on which they are based, therefore the construction of cladistic hypotheses remains the crucial first step. Chapter 3 outlines the most appropriate methodological approaches and examines the role of fossil data in constructing phylogenetic hypotheses. All taxa have to be defined and identified consistently if they are to be useful for evolutionary studies; the way in which taxa are constructed is considered in Chapter 4. As different kinds of groupings carry different implications for evolutionary studies, it is important that we understand the limitations of each.

Species, taxa, and macroevolution

Another major area of palaeontology that burgeoned during the 1970s and 1980s, gaining inspiration and strength from the apparent success of the taxic approach to evolutionary patterns, was the field of macroevolution. Macroevolution covers a huge variety of concepts and processes, and means different things to different workers. For example, Stanley (1979) considered macroevolution to be evolution caused by speciation and selection among species, whereas Valentine (1990) considered all higher taxa (including paraphyletic groups) to be active participants in macroevolutionary processes. In contrast, Levinton (1988, p. 2) took a very restricted view of macroevolution, defining it as 'the sum of those processes that explain the character-state transitions that diagnose evolutionary differences of major taxonomic rank'. In its most generally accepted formulation, macroevolution comprises large-scale patterns of diversification, extinction, and body-plan innovation arising from phenomena and processes working at or above the species level. Species are conceptualized as individuals and as the unit of selection, speciation as the source of variability, and species selection as an equivalent process to natural selection acting on extinction and speciation rates of individual species.

The major difficulty with macroevolution is determining the level in the hierarchy at which selection is acting. There is the distinct possibility that all species and higher taxonomic macroevolutionary patterns are simply epiphenomena, the result of processes acting at some lower level in the hierarchy (Vrba & Eldredge, 1984; Cracraft, 1989; Hoffman, 1989). A great deal of macroevolutionary theory hinges on the nature of species in the fossil record and this forms the subject of Chapter 2. Species are seen as the basic units of evolution and as active participants in a whole host of postulated processes. However, the cladistic revolution has led some to question whether there is any distinction between species and higher taxa (Nelson, 1989b), and molecular data are now demonstrating that many morphologically defined species are divisible into a number of diagnosable and geographically discrete subunits which can be grouped hierarchically on the basis of genetic data (Avise *et al.*, 1987; Avise & Ball, 1990). Since species in the fossil record are groups established on the grounds of morphological similarity, they are subject to the same problems of paraphyly and plesiomorphy as are other taxa. The consequences of this for macroevolutionary theory are discussed in Chapter 4.

2 Species in the fossil record

Species concepts

Many workers consider species as the one category that is real: 'What really evolve are species: they speciate and become transformed. Genera and higher taxa do nothing whatsoever.' (Ghiselin, 1984b, p. 213). They are the smallest unit usually identified in taxonomic studies and are widely viewed as the level in the hierarchy most actively involved in generating macroevolutionary patterns. Yet there are many different species concepts, each carrying its own set of implications and limitations. Species in the fossil record constitute the raw data from which patterns and processes of evolution are deduced. The present chapter therefore sets out to clarify what fossil species represent and what can legitimately be inferred about them.

The literature on species concepts is huge, but basically they are of two kinds: (i) those that emphasize the biological processes that are thought to be involved (process-related definitions); and (ii) those that emphasize the operational means by which they are recognized (pattern-based definitions).

Process-related definitions

Species concepts that rely on knowledge of the process that gave rise to them include the 'biological', the 'evolutionary', and the 'Hennigian' species concepts.

The biological species concept. This has wide support among biologists, and Mayr (1969, 1976, 1982) has been one of its strongest advocates. It is based on the empirical observation that individuals within populations can successfully interbreed, whereas those from different species are in general unable to interbreed. This led to the definition of species as 'groups of interbreeding natural populations which are reproductively isolated from other such groups' (Mayr, 1969, p. 26). Note that morphological disparity has no place in the definition of biological species and its application relies first and foremost on the identification of the actual or inferred reproductive potential between populations. It is interbreeding that gives species their cohesion and distinguishes them from other taxa (Wiley, 1981; Templeton, 1989).

One difficulty with this definition is that hybridization can often be accomplished, under artificial situations, between individuals from different gene pools that would otherwise not normally interbreed.

To overcome this Paterson (1981, 1985) and others (e.g. Templeton, 1989; Turner & Paterson, 1991) have modified the species concept to stress the importance of autorecognition factors in defining species boundaries. Populations from different gene pools that could potentially interbreed to produce viable offspring, but which have evolved different mate-recognition behaviour, are accepted as valid species. Differences in courtship display, for example, can be very important in preventing closely-related and morphologically very similar species from interbreeding. Thus Paterson (1985) defined a species as 'that most inclusive population of individual, biparental organisms which shares a common fertilization system'.

Avise & Ball (1990, pp. 58–59) have expanded on the biological species concept from the view of molecular genetics, recognizing biological species as 'groups of actually or potentially interbreeding populations isolated by intrinsic reproductive barriers'. They also defined subspecies as 'genetically distinct sets of geographical populations exhibiting considerable historical, phylogenetic separation from one another'. Such subspecies are potentially able to interbreed but are sufficiently distinct, genetically, that the same hierarchical pattern linking geographically discrete units is apparent in the phylogenetic trees constructed from more than one gene. In their view subspecies are allopatric populations separated solely by extrinsic gene-flow barriers that have not been established long enough to allow intrinsic reproductive barriers to evolve through genetic divergence.

The evolutionary species concept. Simpson (1961) attempted to apply the biological species concept to encompass fossil species by defining species as 'an ancestral–descendant sequence of populations evolving separately from others and with its own evolutionary role and tendencies'. However, without direct evidence of reproductive potential, Simpson had to assume a direct correlation between reproductive isolation and degree of morphological divergence within lineages to establish species; thus, in effect, he resorted to a morphological definition. Van Valen (1976) adopted a very similar view, defining species as lineages which occupy an adaptive zone minimally different from that of any other lineage and which evolve separately. Again, however, adaptive zones are a matter of inference, and in practice the real key to species recognition is morphological similarity.

The Hennigian species concept. Although Henning (1966) made a major impact on our understanding of phylogeny and systematics through his insight into the nature of higher taxa, his species concept remained firmly based on the biological processes that were supposed to give rise to species. Hennig defined species in very much the same way as Mayr and the neo-Darwinists, identifying reproductive isolation as the key event at which new species come into existence. The difference with Hennig's scheme, however, is that the limits of species are established on the basis of tree topology. Each terminal branch or internode on an evolutionary tree is, by definition, a species. Thus a species ends when, and only when, it splits to give rise to new species or becomes extinct.

This approach is theoretically sound, but operationally very difficult to implement. Under Hennig's model of evolution, a species always splits to give rise to two new species, irrespective of whether new characters evolve in just one or in both of the daughter species. This means that potentially the parental species and one of the daughter species could remain morphologically indistinguishable. On the other hand, a succession of morphological forms would all be placed in the same Hennigian species if they formed a single unbranched lineage through time. Any branch between speciation events represents a single species irrespective of how much morphological change takes place between end members.

A very similar species concept was advocated by Wiley (1978, 1981), but he allowed unequal partitioning of the ancestral species such that a new species might arise through peripheral speciation without the extinction of the parental species.

Problems with process-related species concepts. Any species definition that is derived from a particular view of the evolutionary process is adopted at the expense of practical criteria for their recognition. Using reproductive isolation to define species, for example, makes their identification largely conjectural. There is ample evidence of mismatch between reproductive isolation and morphological distinctness, which is apparent from the literature on cryptic, sibling, and polytypic species (Knowlton, 1993). The wealth of molecular data now becoming available also demonstrates that geographic populations within many biologically reproductive species show hierarchical structure when gene phylogenies are constructed (Avise *et al.*, 1987; Avise, 1989; Avise & Ball, 1990; Harrison, 1991). The greater the agreement among independent gene-trees constructed for the same geographic races, the greater and more effective has been their genetic isolation. The concordance shown by phylogenies derived from different genes ranges from poor to excellent, bridging the gap between population genetics and taxonomic hierarchies, and making hard-and-fast boundaries difficult to draw. Thus the distinction between populations, subspecies, species, and higher taxa becomes a question of relative inclusiveness.

In reality, species have almost always been recognized and differentiated on their diagnosable traits (Sokal & Crovello, 1970). To replace a description of an empirical condition with its explanatory hypothesis is self-defeating, and should be avoided (Brady, 1985). It would therefore be better to base our species concept on an operational definition related to how species are recognized in practice, and to employ a pattern-based species concept.

Pattern-based definitions

Species definitions based on empirical observations of the distribution of phenotypic or genotypic characters among individuals have always provided the only practical method of recognizing species. Again there are several competing species concepts, including the 'phenetic', the 'phylogenetic', and the 'monophyletic' species concepts. These can all generally be grouped under the term 'morphospecies' concepts when based on morphological data.

The phenetic species concept. The phenetic species concept was developed by numerical (phenetic) taxonomists of the 1960s and 1970s. A clear statement of the morphospecies as a phenetic concept is given by Sokal & Crovello (1970) and by Temple (1987). Species are distinguished on phenetic differences, be they morphological, physiological, biochemical, or genetic. All characters that vary among the specimens under study are scored and a taxon–character matrix constructed. This matrix is then subjected to multivariate analysis (usually principal component analysis or discriminant function analysis) in order to identify discrete clusters of individuals based on total morphological variation (reviewed in Sokal, 1986). The smallest clusters recognizable – either by eye or by using a cut-off level – then constitute species.

Morphological analysis can be carried out independently of stratigraphic evidence (the *global phenetic method*). Alternatively, samples can first be grouped stratigraphically and clusters of individuals identified at each stratigraphic level. Clusters are then compared between levels and those deemed sufficiently similar are merged. The latter approach is termed *stratophenetics*.

Budd & Coates (1992) used phenetic methods to divide Cretaceous corals of the genus *Montastraea* into a series of phenetic species. These species were defined solely on a global assessment of phenetic distance without any recourse initially to the stratigraphic record. They began by constructing a taxon–character data matrix. From this, using an average linkage cluster analysis (UPGMA), they produced a dendrogram summarizing overall phenetic similarity. Then clusters were identified by selecting (by convention) a cut-off level on the dendrogram, below which distances were deemed great enough to merit species recognition. From 117 colonies of *Montastraea*, Budd & Coates finally recognized eight species-level clusters. However, 35 colonies did not always appear in the same cluster when different clustering parameters were applied, and these were initially left unclassified.

Budd & Coates (1992) also applied the stratophenetic approach to their data by splitting the samples into four geological periods. By comparing samples from different stratigraphic levels they were able to discriminate 16 clusters in total, which were taken to represent species-level groups (Fig. 2.1).

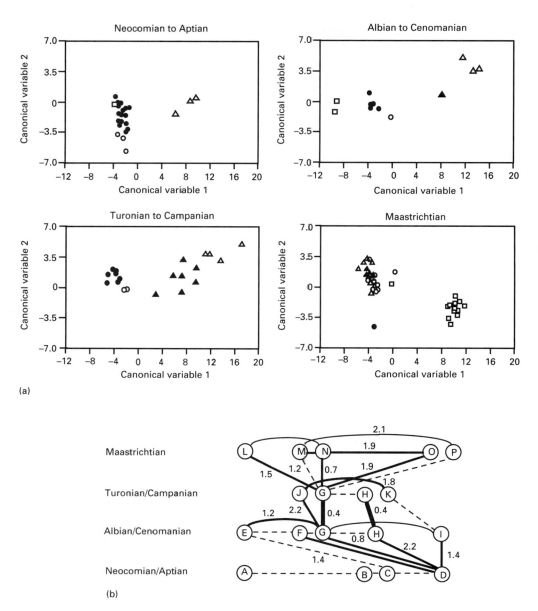

Fig. 2.1 Results of a multivariate morphometric analysis of Cretaceous *Montastrea*-like corals (Budd & Coates, 1992). A data matrix was compiled using 10 linear distances measured on 117 colonies. UPGMA average linkage cluster analysis identified eight groups. (a) A canonical discriminant analysis was then run on these eight groups after splitting the data into four stratigraphic periods (upper diagrams). Four clusters are apparent in the Neocomian–Aptian and Turonian–Campanian assemblages and five in the Albian–Cenomanian and Maastrichtian assemblages (indicated by different symbols). (b) The scores derived from the canonical discriminant analysis were then compared between clusters at different levels and only two were found to be sufficiently similar to justify merging (taxa G and H). Thus a total of 16 species were recognized on the basis of this analysis (taxa A–P in (b)). These species were then linked using the shortest Manhattan distance between pairs, to construct a phenetic network (b).

Phenetic methods have also been applied recently to fossil bryozoans by Cheetham (1986, 1987), Cheetham & Hayek (1988), and Jackson & Cheetham (1990). Cheetham (1986, 1987) constructed a taxon-character matrix for populations of the ascophoran bryozoan *Metrarabdotos* from the Upper Oligocene–Lower Pliocene of the Dominican Republic. These showed considerable morphological complexity and Cheetham was able to identify 46 characters that could be scored, 20 of which were continuous variables, while the rest were discrete variants. Then, taking sample populations from discrete stratigraphic levels, Cheetham ran a series of discriminant function analyses, based on all canonical scores, to identify phenetic clusters of individuals. These clusters were then linked between stratigraphic levels according to their perceived phenetic similarity and stratigraphic proximity, and morphological discontinuity was used to define species boundaries (Fig. 2.2).

Occasionally, with detailed sampling and morphometric work, patterns of morphological change can be mapped out in single lineages. Geary (1990, 1992) carried out such a study on the brackish to fresh-

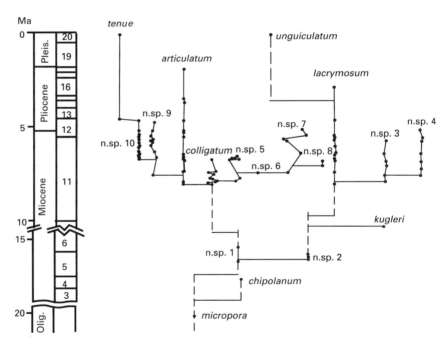

Fig. 2.2 A stratophenetic tree for species of the Neogene bryozoan *Metrarabdotos* (Cheetham, 1986). This is based on 46 morphometric characters, weighted using a series of discriminant analyses and then used to group samples by applying 2 UPGMA cluster analysis. This identified 18 clusters (species) which were then grouped into nearest neighbour clusters using 'overall morphological similarity and stratigraphic position'. Note, however, that the relative weighting given to these two lines of evidence is apparently at the discretion of the taxonomist and proceeds on an *ad hoc* basis. Unlike the example in Fig. 2.1, the resultant tree does not match the phenetic network. n.sp., New species.

water gastropod genus *Melanopsis* from the middle to late Miocene
Paratethyan sequences of the Pannonian basin (eastern Europe). She
was able to demonstrate, using multivariate techniques (discriminant
analysis), that over a period of about 2 Ma there was a long-term persistence and ultimate disappearance of intermediates between two
morphologically discrete end members (Fig. 2.3), *Melanopsis vindobonensis* and *M. fossilis*. By outgroup comparison with a third species,
M. impressa, she demonstrated that these two species are sister groups
that appear in the fossil record in Pannonian stage C, and coexist to
the end of that period. This pattern represents, however, not simply
speciation through gradual differentiation, but a much more complex
process. This can be seen from the fact that at least one stratigraphically early sample shows complete differentiation of *M. vindobonensis*
in the absence of *M. fossilis*. Geary was uncertain whether to interpret
this evolutionary pattern as ecophenotypic variation within a single
biological species or as interspecific hybridization. Thus, even in this
well documented case, the question of when or if reproductive isolation arose in the fossil record remains unanswerable.

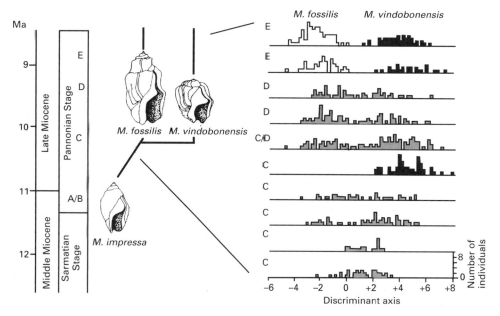

Fig. 2.3 Morphological divergence in the late Miocene gastropod *Melanopsis* (Geary,
1992). Two morphological species of *Melanopsis* are recognized in the Pannonian
basin of central Europe, *M. fossilis* and *M. vindobonensis*. These apparently arose
from an earlier species *M. impressa* during Pannonian stage C (left). Twelve
morphological characters were scored in a large number of individuals from 10
localities ranging in age from Pannonian stages C–E. These were then used in a
discriminant analysis and plots of the first canonical discriminant axis for each
population (right). By late Pannonian stage E the two species are clearly distinct. At
lower levels the two species intergrade, except at one locality (Hennersdorf), where
only *M. vindobonensis* morphologies are found.

Although phenetic methods can work extremely well for morpho-species discrimination, and can often give a more effective summary of total morphological variability than any univariate or bivariate plot, they also have some drawbacks. The identification of species as all individuals that cluster within a certain area in a multivariate analysis often makes practical diagnosis difficult other than by phenetic methods. The major axes in a discriminant function analysis or principal components analysis cannot be linked to observable character differences directly because they are constructed from total variance across all characters. In most cases only one or two characters dominate each canonical discriminant function (CV), but all characters have some input into each CV, and the same character may contribute significantly to more than one CV. Thus, interpreting precisely what discriminates species from the morphological complex of features fed into the analysis is not always simple. Typically, after multivariate analysis has been used to identify clusters, it is necessary to define the species on tangible morphological differences in specific characters. For example, Cooper & Ni (1986) found principal coordinates analysis very useful for discriminating between pairs of closely related species of the early Ordovician graptolite *Pseudisograptus*. But, having identified the clusters, they then constructed diagnoses for the species on discrete character differences.

More importantly, defining species using multivariate techniques limits any subsequent analysis of species relationships to phenetic methods which, as discussed below, are inferior to parsimony. For those interested, Temple (1980) provides a very clear exposition of the global phenetic approach to phylogeny reconstruction, and the strato-phenetic method has been used by Gingerich (1979) and Fordham (1986) for establishing phylogenies. Cheetham & Hayek (1988) and Budd & Coates (1992) used scores derived from a canonical discriminant analysis as characters for parsimony analysis. However, this use of transformed character data seems highly questionable. Not only are the characters continuous variables that have to be divided arbitrarily, but they are also not necessarily independent of one another.

The phylogenetic species concept. Cladists such as Eldredge & Cracraft (1980), Nelson & Platnick (1981), Cracraft (1983), Schoch (1986), and Nixon & Wheeler (1990) have all advocated that species should be defined as 'the smallest diagnosable cluster of self-perpetuating organisms that have unique sets of characters' (Nelson & Platnick, 1981, p. 12). Cracraft (1983) called this the 'phylogenetic species concept'.

Phylogenetic species are established by comparative analysis, which begins by assembling all consistent, objectively definable characters that can be recognized for the set of individuals under consideration. These individuals are then grouped into clusters on the basis of shared similarity of characters and observed (or inferred in the case of fossil taxa) reproductive continuity within clusters. The latter is necessary to prevent individuals from different sexes and different stages of the life cycle being separated by mistake. The clusters arising from this process are therefore groups of populations which cannot be further subdivided on the assembled evidence, and each cluster is distinguished from all other clusters by at least one unique character state. These clusters constitute the phylogenetic species.

On subsequent cladistic analysis, phylogenetic species inevitably comprise two distinct kinds of groups: (i) minimal monophyletic groups, which are clusters distinguished by the presence of at least one derived character; and (ii) their plesiomorphic sister taxa, which have no diagnostic apomorphies. As pointed out by Eldredge & Cracraft (1980, p. 90) and de Queiroz & Donoghue (1990b, p. 88), the latter are not necessarily monophyletic since they turn out to have been erected on the basis of wholly plesiomorphic traits.

The monophyletic species concept. This concept defines species as the smallest (least inclusive) monophyletic groups recognizable. In operational terms, monophyletic species are discovered in exactly the same way as phylogenetic species, except that only those clusters with at least one defining apomorphy qualify for recognition. This approach has been advocated by Rosen (1979), Donoghue (1985), Mischler & Brandon (1987), and de Queiroz & Donoghue (1988, 1990a,b), but has been criticized on practical and theoretical grounds (Nelson, 1989a; Nixon & Wheeler, 1990; Wheeler & Nixon, 1990).

The advantage of using a monophyletic species concept is that species are defined and recognized in exactly the same way as any higher taxon. By treating species as simply the smallest monophyletic unit in a hierarchy, they lose any claim to special status as the basic unit of evolution and become no different from other taxa (Nelson 1989b, p. 61). Only their level of inclusiveness serves to distinguish species from higher taxa.

One obvious disadvantage is that only some groups diagnosable as morphologically distinct will possess identifying apomorphies, leaving the rest in limbo. To resolve this problem Donoghue (1985) suggested the term *metaspecies* for groups of populations that are morphologically diagnosable but which lack apomorphies. Later Gauthier *et al.* (1988) expanded the concept to include any taxon of unresolved status as a *metataxon*.

Another problem is that the phylogenetic status of a taxon changes through time. Consider a new character evolving in a population. This acquisition of an apomorphy marks those displaying it as a monophyletic group and, being the smallest diagnosable unit, they by definition form a species (Fig. 2.4). If members of this species later acquire another novel character through divergent evolution, this also enables us to recognize and define a new species on the strength of that second apomorphy. The two species become sister taxa on a cladogram. However, the original apomorphy, which defined the parental species

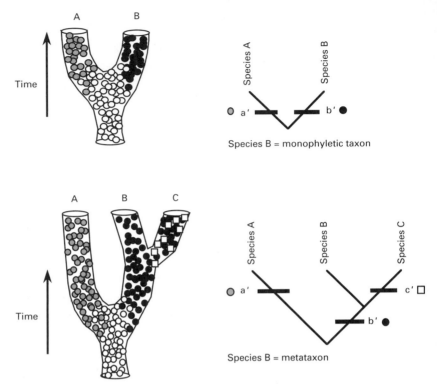

Fig. 2.4 Graphic representation of evolutionary trees (left), and derived cladograms (right) illustrating the differences between metataxa and monophyletic species. (a) Species A and B each have their own autapomorphic character (a' and b', respectively) and are thus monophyletic species. (b) If after some period of time a third species arises which can be recognized by its apomorphic character c', then species B is transformed from a monophyletic taxon into a metataxon.

as monophyletic, now becomes an apomorphy for the clade formed by the parental and daughter species and the parental species changes its status to become a metataxon. The monophyletic species concept thus uses definitions that are time-dependent, which may have disadvantages; for one thing, it means that all present-day monophyletic species are potentially metataxa.

Nixon & Wheeler (1990, p. 217) further criticized the monophyletic species concept, stating that 'species based on topological (cladogram) knowledge are problematic, because they cannot be implemented prior to cladistic analysis.' However, all empirically based species concepts must rely on some form of analytical procedure in order to recognize clusters of individuals, including the phylogenetic species concept.

Species in palaeontology

Since the fossil record provides no access to genetic or ecological information other than by inference, there is no option but to define palaeontological species in terms of observed morphology. Otherwise we would be replacing an empirically derived concept of species with mere speculation. Palaeontological species are thus minimal morphological clusters of individuals deemed useful to establish. But which of the various operational definitions provides the best criteria for species recognition?

Although the phenetic approach offers an excellent method for objectively identifying clusters of individuals, such clusters are only useful if non-arbitrary characters can be found to differentiate them. Species defined using phenetic techniques need to be identifiable on practical grounds and therefore need to be delimited on the basis of specific character differences. Phenetic techniques alone do not seem appropriate, but they do provide a useful first step in determining the smallest diagnosable clusters of individuals, which can then be compared for meaningful character differences.

Both the monophyletic and phylogenetic species concepts employ characters to differentiate minimal morphological units. The implementation of a monophyletic species concept is secondary to the discovery of phylogenetic species: phylogenetic species must be defined first before monophyletic species can be distinguished from among them by cladistic analysis. Should the term species therefore be applied to the smallest monophyletic group or the smallest diagnosable group? The answer depends largely on whether the smallest diagnosable plesiomorphic groups are truly indivisible. If each represents a highly cohesive population with extensive gene-flow amongst its members, there will be no internal hierarchical structure and subdivision will be impossible. In such a case the phylogenetic species concept seems the correct one to use.

If, however, plesiomorphic basal taxa are simply units lying at the limits of resolution afforded by the data to hand, then it is likely that each could be further subdivided hierarchically, given access to additional morphological or genetic information. If this is true, there is no justification for recognizing plesiomorphic basal taxa as formal taxa, any more than there is justification for formally recognizing higher plesiomorphic taxa such as reptiles or invertebrates.

Are basal taxa in the fossil record likely to be truly indivisible, or have most been established by default simply because of the limits of available morphological evidence? My firm conviction is that fossil species are established on incomplete information and represent groups whose boundaries arise because of the practical limitations faced by systematists, not because of any inherent phylogenetic indivisibility of such units. Fossils must always be anatomically less complete than any modern species because soft-tissue, biochemical, and genetic characters are almost never preserved. Furthermore, even considering skeletal anatomy alone, fossils rarely retain the full suite of characters seen in modern taxa. Fossil echinoids, for example, are hardly ever preserved with their suite of complex pedicellariae, and gut wall spiculation has never been preserved, yet both pedicellariae and gut spicules provide a rich suite of skeletal characters used in the classification of extant members. Similarly, how might some basal taxa of trilobite be viewed if we had knowledge of the minutiae of limb structure? It would therefore be very surprising if basal taxa in the fossil record were not delimited for practical reasons, because of limited access to characters. Furthermore, given the congruent hierarchical structure in the gene trees of many demographically dispersed morphospecies (e.g. Wilson *et al.*, 1985; Avise *et al.*, 1987; Avise, 1989; Avise & Ball, 1990; Knowlton, 1993), even the most exquisitely preserved taxa may still encompass a series of hierarchically structured populations and sibling species, since we have no access to genetic data.

Minimal morphologically diagnosable groups in the fossil record therefore comprise a mixture of basal monophyletic taxa (species) and plesiomorphic grades awaiting additional information or higher resolution studies (metaspecies).

Are species different from other taxa?

There has been much discussion as to whether species are any different from other taxa (see Nelson, 1989b). To a large extent the answer depends on how one chooses to define species. If species are defined as interbreeding populations, higher taxa are clearly different. Species defined in terms of reproductive cohesiveness become segments of an unbroken genealogical nexus and consequently their boundaries are constantly in flux. Such units are entirely nominal since they have no spatiotemporal restrictedness (Rieppel, 1988). Taxa, on the other hand, are firmly anchored in time by the appearance of apomorphies and are spatiotemporally bounded entities.

Species have in the past been thought of both as active participants in evolutionary processes and as spatiotemporally bounded individuals. But if they are participants in evolutionary processes, they cannot be spatiotemporally bounded, because evolutionary processes require continuity. Alternatively, if they are spatiotemporally bounded, they are taxa and the historical products of evolutionary processes, not the effectors of such processes. Rieppel (1988) in particular has clarified the distinction between these two views of species.

The crucial question is whether species *as recognized in the fossil record* are different from other taxa. This is important because many current macroevolutionary theories depend on species in the fossil record being different from other taxa and immune to the problems of paraphyly (e.g. Eldredge & Cracraft, 1980; Ax, 1987). They must be participants in evolutionary processes.

From the above it should be apparent that fossil species are taxa. They must be discovered by the application of comparative methods of clustering and differentiation. If we examine minimal morphological

units in the fossil record that systematists refer to as species, they fall
into two categories: (i) those based on single populations from one
locality which can safely be assumed to be interbreeding populations;
and (ii) those based on groups of populations from different localities/
horizons that have been grouped because of observed similarity.

A population of conspecific individuals from a single locality has
cohesion due to interbreeding. Single populations thus display 'tokogenetic relationships', i.e. reticulate branching patterns (Hennig, 1966;
de Queiroz & Donoghue, 1988; Nixon & Wheeler, 1990). The reality of
gene flow among individuals can be verified for extant populations but
can only be assumed for fossil populations. Single locality/horizon
populations in the fossil record are almost certainly indivisible, reproductively cohesive units, but have no spatiotemporal extent.

Populations from different localities or stratigraphic horizons that
cannot be differentiated on morphological grounds do not have the
same claim to biological species status. Cohesiveness due to interbreeding and gene flow is not an issue and populations can only be
grouped on the basis of observed morphological similarity. Here grouping of populations is no different from grouping species or any other
higher taxa, and there is no fundamental distinction between the way
species and higher taxa are recognized.

Thus, as soon as we move from inferring tokogenetic relationships
within a single population to constructing hypotheses of relationship
through grouping populations, we are discovering taxa and the concepts of paraphyly and monophyly are just as applicable at this level as
they are for higher taxa. The fundamental split then lies between
single populations and groups of spatiotemporally dispersed populations, with the former probably representing biologically cohesive
participants in evolutionary processes, and the latter representing taxa,
the historical by-products of evolutionary processes.

De Queiroz & Donoghue (1988, p. 334) suggested that it might be
more appropriate to abandon the term species, or simply restrict it to
designate 'interbreeding groups', leaving monophyletic groups free of
any association with the biological species concept. On the other
hand, species were conceived as typological constructs, and have been
used as the lowest units in the hierarchical classificatory system since
the time of Linnaeus. It was only with the Darwinian revolution that
species came to be thought of as interbreeding groups and as active
participants in evolutionary processes. It could therefore equally be
argued that species should return to being basal taxa, freed of any
process-related implications that they have only relatively recently
acquired. Whatever the case it is clear that the two notions of species
must be kept distinct to avoid inappropriate conflation.

Throughout this text I shall refer to minimum morphological units
in the fossil record that are consistently diagnosable on the basis of
meaningful characters as *phena* (singular phenon). This term is already
established in the literature (e.g. Mayr, 1969; Fordham, 1986), though

it has usually been used for clusters whose boundaries are gradational. Phena are the basal units that systematists recognize, and it is only by subsequent morphometric and cladistic analysis that monophyletic species taxa are distinguished from among them. They more or less approximate to phylogenetic species in concept, but it is not always clear what phena actually represent because sexual dimorphism and variation during life stages can only be inferred and not directly observed in the fossil record. Morphometric analysis is required to ensure that phena do not simply represent different sexes or stages in the life cycle of taxon, and cladistic analysis is required to establish which phena are monophyletic taxa (species) and which are grades (metaspecies). Although all phena should be named, there should be some convention by which species taxa are differentiated from metaspecies grades in the Linnaean hierarchy.

How phena are recognized

Every phenon has to accommodate a certain amount of morphological variation, unless based on a unique specimen. No two individuals are ever identical, and the situation is made more difficult by ontogenetic change, sexual dimorphism, and polymorphism inevitably found within actively interbreeding populations. Differences in the style and quality of preservation further complicate matters. Thus, it is not a simple task to infer the range of morphological variation among individuals thought to compose a cohesive population.

Phenon recognition is basically a two-step process. The first step is *to document the range of variation seen within single locality/horizon populations*. This is essential not only to establish the range of phenotypic variability among individuals of the same size and stage, but also to identify morphological changes that occur during ontogeny, and for the recognition of sexual dimorphism and other polymorphisms that may be found within the population. If this is not done, individuals from single reproductive communities may be spuriously subdivided on attributes specific to one sex or to stages in the life-cycle. Empirical observation of present-day populations makes it patently clear that we must define our basic taxonomic unit above the level of variation encountered among members of fully interbreeding populations.

This assessment of intra-population variation is best accomplished by the application of biometric techniques. Thus ontogenetic variation in characters is commonly investigated by the use of bivariate plots in which one axis is a measure of overall size (Fig. 2.5). The recognition of sexual dimorphism is more difficult and is characterized by a common ontogenetic trajectory in pre-adult forms which later diverges to give rise to distinct adult morphs (Fig. 2.6). This is further discussed below.

Intra-population studies are also needed to identify fabrication noise

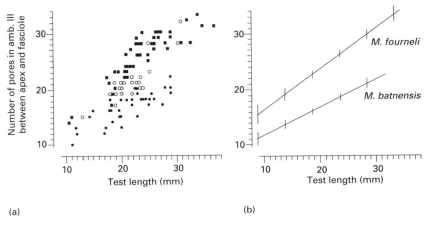

Fig. 2.5 Morphometric discrimination between two closely related phena of the Upper Cretaceous echinoid *Macraster* (Smith & Bengtson, 1992). Samples were collected at five stratigraphic levels from late Cenomanian to early Coniacian in age. The late Cenomanian and early Turonian populations could not be differentiated statistically, nor could the late Turonian and early Coniacian populations. The density of pore-pairs in the frontal ambulacrum appeared to give the best discrimination between populations, but varied with test size. (a) The bivariate plot of test size against number of ambulacral pores discriminated the two end members clearly, but the mid-Turonian sample was intermediate. ●, late Cenomanian and early Turonian populations; ○, mid Turonian population; ■, late Turonian and early Coniacian populations. (b) Regression analysis for the two end-member populations (bars represent 2 standard errors).

created by disease and/or abnormal development. For example, Roth (1989) showed that elephant tooth structure is particularly prone to developmental abnormality and suggested that such fabrication noise may have led to a gross overestimate of the number of fossil elephant species.

Having established the range of morphological variability in a single population, the next step is *to compare populations from different localities/horizons to discover whether significant morphological differences can be discerned*. Populations with unique character states are separated, while those that show no discernible differences in character states are aggregated until all populations are assigned to diagnosable phena. Again, clustering can often be aided by the use of biometric techniques and good examples of the use of morphometric techniques in species discrimination can be found in Temple (1987 – early Silurian brachiopods) and Hohenegger & Tatzreiter (1992 – Triassic species of the ammonoid *Balatonites*).

At this stage taxonomic standardization is vital, so that taxa are consistently defined and recognized. Major discrepancies in the usage of taxonomic names by different authorities commonly exist in the literature (Culver *et al.*, 1987). McCune (1986), for example, reexamined species of *Semionotus*, a group of actinopterygian fishes found

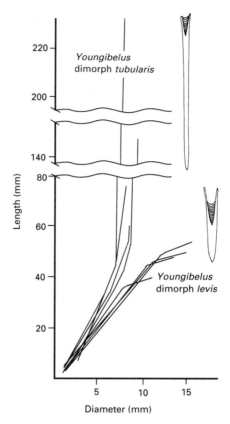

Fig. 2.6 Growth curves for two co-occurring belemnite phena *Youngibelus tubularis* and *Y. levis* from the Lower Toarcian (Lower Jurassic) of Britain (Doyle, 1985). The two species are found in equal abundance and are interpreted as sexual dimorphs. The initial stages of growth in the two forms are indistinguishable and it is only in the adult stages that the two forms diverge dramatically.

in the Triassic and Jurassic. Of the 42 nominal species assigned to this genus, she could find justification for only three morphologically diagnosable phena, plus a fourth that had previously been classified in a different genus. Many of the older species had been based on poorly preserved material or had been misclassified.

Examples in practice

The following three cases exemplify some of the approaches and problems associated with phenon discrimination. The first involves simple population discrimination in graptolites; the second deals with the problems of discriminating phena within a more or less continuous lineage of Silurian brachiopods; and the third discusses the recognition of sexual dimorphism in ammonites and how phena may be grouped into species taxa.

Samples from a single locality/horizon

The Arenig (Lower Ordovician) graptolite *Pseudisograptus* displays a large variation in morphology and has consequently proved a difficult taxon to split up systematically. Cooper & Ni (1986) used biometric analysis of populations from single localities to revise the nominal species of *Pseudisograptus* and establish phena. Their analysis concentrated on the proximal portion of graptolite rhabdosomes, since this had previously been shown to possess the most taxonomically useful features in graptolites (Cooper & Fortey, 1983). Various measurements were taken from this region and a number of discrete character states were also employed, such as the outline form of the thecal chambers and the mode of development and addition of theca during early astogeny (colony ontogeny).

Cooper & Ni then used the biometric data to construct bivariate and univariate plots. This allowed a simple test of whether or not samples from a single locality/horizon could be considered to belong to a single phenon; disjunct distribution was used as evidence for separate phena. In some cases this provided unambiguous separation of phena (Fig. 2.7); in others the situation was less clear-cut and some overlap occurred creating bimodal distributions. In the latter cases, Cooper & Ni measured total morphology of populations using the multivariate technique of principal component analysis (PCA). In order to remove the biases that arise because of the various scales used for different measurements, they transformed their data matrix into a correlation matrix. From this they obtained almost completely non-overlapping differentiation of four phena within the *P. manubriatus* group and clear differentiation amongst other phena (Fig. 2.7). However, since the characters used were continuous variables and single character analysis was unable to distinguish these four clusters of *P. manubriatus* unambiguously, Cooper & Ni chose to recognize them at subspecific level.

This example demonstrates that phena can range from clearly objectified clusters diagnosable on the basis of single character states, to more-or-less arbitrary divisions of continuous variability, established using multivariate techniques and undefinable in terms of single character differences. Phena whose boundaries are imposed arbitrarily may still serve a useful purpose (as in the following example) but they do not merit recognition as formal taxa because their composition is set by convention. Phena with diagnostic characters, on the other hand, represent groups which can be objectively discovered and are worth naming. Only further analysis will tell whether these are monophyletic species taxa or represent metaspecies grades.

Phena within a single evolving lineage

One of the most stratigraphically useful benthic marine invertebrates of the Silurian is the brachiopod *Stricklandia*. This brachiopod is

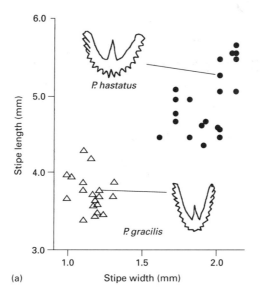

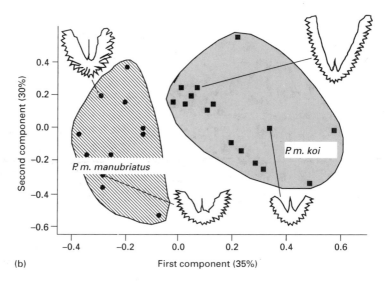

Fig. 2.7 Biometric discrimination of phena amongst the Lower Ordovician (Arenig) graptolite *Pseudisograptus* (Cooper & Ni, 1986). (a) Simple bivariate plot of stipe length against stipe width for 39 specimens, which allows clear differentiation of the two phena, *P. hastatus* and *P. gracilis*. (b) First two axes of a principal component analysis based on five characters, equally weighted, for all short siculate forms from the same locality and horizon. This was used to differentiate between two named subspecies, *P. manubriatus manubriatus* and *P. manubriatus koi*.

extensively used for biostratigraphic dating of Lower Silurian rocks in North America, the British Isles, eastern Europe, and Scandinavia. Although external features (e.g. shell shape, ornamentation and length of the interarea) are highly variable within single populations, the internal structure of the cardinalia (Fig. 2.8) has proved taxonomically significant and has allowed workers to recognize five phena, variously assigned subspecific or specific status.

Baarli (1986) undertook a detailed biometric analysis of this lineage in Norway in order to place the systematics of the group on a secure foundation. *Stricklandia* occurs at various levels within approximately 235 m of section in the Llandovery of the Asker area, Norway. Baarli collected samples at 2.5–5-m intervals throughout the section and extracted *Stricklandia* valves. Although sample size was generally small (ranging from 5 to 67), stratigraphic sampling was relatively dense (391 individuals from 32 levels). A variety of measurements were then made on the cardinalia, of which the most important proved to be: (a) the length from the posterior point of the cardinalia to the anterior point of the outer plates; (b) the length from the posterior point of the cardinalia to the anterior point of the inner plates, where they are fused with the brachial processes; and (c) the height measured from the anterior point of the inner plates where they are fused with the brachial processes vertically down to the base of the valve floor. In order to standardize against variation in size, measurements were

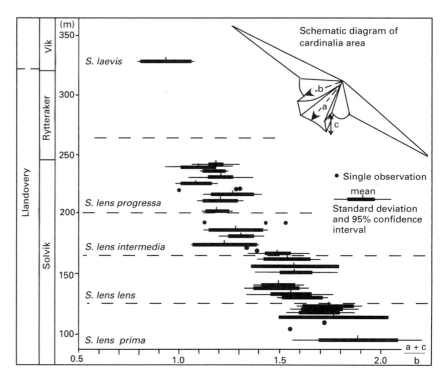

Fig. 2.8 Biometric analysis of 391 specimens of the early Silurian brachiopod *Stricklandia* from 32 sampled levels in the Asker region of Norway (Baarli, 1986). The plot given is of the form index (a+c)/b taken from the cardinalia structure on the interior of the valve (see insert top right). Nominal species and subspecies recognized by Baarli are indicated.

expressed as a ratio of length (b), since this appeared relatively in variable and a suitable measure of overall size. In addition to the Norwegian material, Baarli also examined a large number of specimens from Estonia.

Baarli found end members significantly different for four of the seven indices used. The most useful index (i.e. the one with the most discriminatory power) was (a+c)/b. In addition to strictly biometric measurements, some qualitative observations on muscle-scar development and the degree of extension of outer plates beyond their fusion with the brachial process were needed to aid discrimination.

Despite the obvious differentiation between end members of this lineage, clear phenon boundaries between named subspecies are lacking. *Stricklandia lens lens* and *S. lens intermedia* show the clearest morphological break, but, by comparison with regions outside the Asker area, where the transition is more gradual, Baarli argued that this apparent morphological discontinuity is due to a local diastem in the section. Given a sample of reasonable size, it is obviously possible to place individual populations within very limited stratigraphic bands in the section. Even better resolution might be achieved with a multivariate approach using more characters (e.g. muscle-scar data).

The phena identified by Baarli are established purely by convention – provided other workers know and apply the same convention (i.e. use the same arbitrary cut-off points along a continuous variable to separate phena), these phena can be consistently recognized. In this case it is clear that nominal phena can provide very high resolution stratigraphic subdivision applicable over a wide geographic area. Nevertheless, such phena are purely notional and only one monophyletic species taxon – or a suite of successively less inclusive taxa – can be objectively justified in terms of characters. Named phena serve no real purpose, because once the convention is known the highest stratigraphic precision comes from applying the morphometric criterion directly on populations. Determining stratigraphic level by placing sample populations into named phena can only result in less precision.

In this case, therefore, phena are entirely nominal and characters provide no means by which taxa can be differentiated except by convention. Objective boundaries cannot be drawn and it would seem best therefore to treat the entire lineage as the smallest indivisible species taxon on current evidence.

Sexual dimorphism in ammonites

Sexual dimorphism is a common feature in several marine invertebrates (e.g. Westermann, 1969; Calloman, 1981; Lehmann, 1981; Smith, 1984b; Doyle, 1985; Howarth, 1991). Where dimorphism is pronounced, as in insects or ammonites, there is a danger that sexual dimorphs will be classified as different species. Arkell (1957) includes many examples where the possession of lappets or 'gerontic simple ribs', or simply size – all of which are sex-linked characters – were used to set up subgenera or genera. In recent years there has been much more awareness of the problems of sexual dimorphism in ammonoids and the widespread recognition of macroconch (?female) and microconch (?male) dimorphs (e.g. Calloman, 1981; Lehmann, 1981).

An exemplary study of sexual dimorphism is provided by Howarth (1991) in his revision of the Toarcian (Lower Jurassic) ammonite family Hildoceratidae. By careful field collecting from single beds or nodule horizons Howarth was able to study associated macroconchs and microconchs and provide a detailed documentation of sexual traits.

The first step in recognizing sexual dimorphism is to determine whether individuals in a sample are adult and fully mature. There are several characters that can be used in the hildoceratids: (i) modification of the growth of the body chamber near the mouth border, which is usually accomplished by a change in the spiral pathways of both outer and inner seams, increasing the umbilical diameter and sometimes decreasing the whorl height; (ii) modification of ribs in the final part of the body chamber; (iii) development of lateral lappets or a constriction in the mouth border, when these features are absent in earlier stages; and (iv) crowding of the last 2–4 suture-lines.

Having identified the traits specific to adults, the next step is to compare the morphologies of individuals prior to the onset of sexual dimorphism. Landman (1989), for example, used bivariate plots of umbilical diameter versus shell diameter to compare the ontogenies of co-occurring phena of Upper Cretaceous scaphitids (Fig. 2.9). He found remarkably good correspondence in growth trajectories for all such pairs.

Howarth (1991) found that the macroconch:microconch ratio was extremely variable. The best sampled species, *Eleganticeras elegantulum* and *Cleviceras exaratum*, had macroconch:microconch ratios of 1.38 and 0.87, respectively. However, other species departed significantly from this: in *Harpoceras soloniacense* the ratio is 8:1, while in *H. serpentinum* only 1 in 300 specimens is a microconch. Howarth attributed this variation to single-sex shoaling, post-mortem sorting, differential destruction of shells during burial and diagenesis, and collection bias. For example, *Tiltonoceras* is found in only one horizon in Britain, at the top of the Marlstone Rock Bed in Leicestershire, where the only complete individuals are small. Larger specimens that may have been macroconchs are represented at this level only by fragments.

The size difference between dimorphic pairs can be enormous; Howarth (1991) found ratios generally between 4:1 and 7:1, based on large samplesizes. Individual variation within each dimorphic form is also considerable, with a ratio of largest to smallest for complete adults of 2:1 and 3:1 in macroconchs and microconchs, respectively.

Landman (1989) undertook a detailed investigation of co-occurring

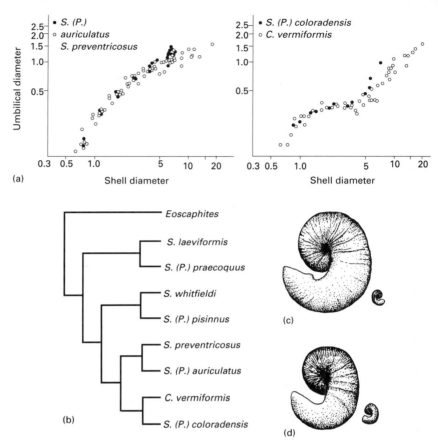

Fig. 2.9 Dimorphism in late Cretaceous scaphitid ammonites (Landman, 1989).
(a) Growth trajectories of individual co-occurring pairs of pteroscaphite and
clioscaphite ammonites plotted on a logarithmic scale. Note how the growth curves
for pairs of ammonites closely match one another but are distinct from those of
other pairs. (b) Cladogram based on an assessment of 10 characters. This suggests
that each pteroscaphite pairs with its cooccurring clioscaphite. (c) *Scaphites
(Pteroscaphites) auriculatus* (small pteroscaphitid) and *Scaphites (Scaphites)
preventricosus* (large clioscaphitid). (d) *Scaphites (Pteroscaphites) coloradensis*
(small) and *Clioscaphites vermiformis* (large).

phena of Upper Cretaceous scaphitid ammonites, including a cladistic
analysis. Adult scaphitids are easily recognized because they show
strong uncoiling of the body chamber. In the Western Interior, USA
during the Turonian to Santonian, two distinct groups occur: a group
of small species with apertural projections classified as *Scaphites
(Pteroscaphites)* (pteroscaphitids) and much larger species classified
as *Scaphites (scaphites)* or *Clioscaphites (clioscaphitids)*. These two
forms co-occur for approximately 7 Ma and appear to have evolved in
parallel. Landman was able to demonstrate that the ontogeny of co-
occurring pairs prior to sexual maturity of the pteroscaphitid was
more similar between the pairs of co-occurring pteroscaphitids and
clioscaphitids than between species assigned to the same genus. Thus,

co-occurring pteroscaphitid and clioscaphitid morphs all appeared as sister taxa in the cladogram (Fig. 2.9).

Landman rejected sexual dimorphism as an explanation for the observed sister group pairing of pteroscaphitid and clioscaphitid phena. Instead he postulated that each pteroscaphitid arose from its sister clioscaphitid by speciation, through the process of progenesis (precocious sexual maturation). He rejected sexual dimorphism because of the following observations:

1 Presumed sexual dimorphism had been recognized within each pteroscaphitid and clioscaphitid (e.g. Cobban, 1969).
2 Variation in size and shape between dimorphs is much greater than previously reported.
3 The size ranges of microconchs and macroconchs do not overlap.
4 Pteroscaphitids are not always found with clioscaphitids.

However, in the light of Howarth's (1991) findings, sexual dimorphism cannot be dismissed so easily. The sexual dimorphism recognized by Cobban (1969) and referred to by Landman is well within the range of size and morphological variation found by Howarth within macroconchs or microconchs of single populations, and in many of Howarth's dimorphic pairs there was no overlap in size of the adult microconchs and macroconchs. Finally, co-occurrence is indicative of, but not necessary for, the recognition of sexual dimorphism, since preservational and biological factors can result in beds containing only microconchs or only macroconchs. Thus there is simply no need to consider multiple convergent evolution through progenesis in pteroscaphitids.

In conclusion, the case for pteroscaphitid and clioscaphitid phena as sexual dimorphs seems extremely strong, and, as the extent of sexual dimorphism in ammonites becomes better established, more such taxonomic revision will surely be necessary. Landman's (1989) cladistic analysis is both pioneering and reassuring because it identifies sexual dimorphs by pairing of macroconch and microconch forms. Since each pteroscaphitid–clioscaphitid pair is supported by at least one derived character, they can all be considered monophyletic species, not meta-species grades. The application of cladistics to ammonite systematics is long overdue.

Summary

In this chapter I have tried to explain why it is best to view species in the fossil record as taxa, created as a by-product of evolutionary processes and recognized by the application of comparative methods. Given the nature of fossils, there is no practical alternative but to adopt an empirical, morphology-based definition of species. Pattern-based species concepts require: (i) that species are the minimal clusters of individual organisms diagnosable on the basis of observable characters; and (ii) that such clusters should include all forms that represent different stages, sexes, or castes of the same reproductively unified population. The latter can be directly observed for extant organisms, but must be indirectly inferred for fossil taxa. Consequently the certainty with which pattern-based species can be recognized in the fossil record is less than that for extant organisms.

In practice, minimally diagnostic morphological groups – here referred to as phena – range from nominal clusters whose boundaries are set by convention, to clearly discrete clusters with their own set of diagnostic characters. Since only objectified groups should be considered as taxa, only the latter should be formally recognized and named.

Finally, the monophyletic species concept is considered the most appropriate to employ in the fossil record. As fossils are anatomically incomplete, often grossly so, it seems very unlikely that the smallest recognizable plesiomorphic taxa represent truly indivisible entities. Therefore they must represent grades which are potentially subdivisible hierarchically, given access to more complete information. Their boundaries are set by the limitations of taxonomic practice rather than any inherent biological factor. Minimal diagnosable morphological clusters in the fossil record are therefore either monophyletic clades or plesiomorphic grades. The former are species taxa; the latter are metaspecies.

3 Parsimony, phylogenetic analysis, and fossils

The cladistic revolution

Throughout the 1970s and early 1980s fierce debate reigned amongst systematists concerning the relative merits of different methodological approaches for classification and phylogeny reconstruction. Hull (1970) recognized three distinct schools involved in this dispute: the cladists or phylogeneticists, the pheneticists, and the evolutionary systematists (an excellent historical review of the methodological stance of each is given by Schoch, 1986). As the dust settled, it became clear that there were in fact only two different methodological approaches to phylogenetic reconstruction: (i) those that constructed relationships on the basis of only the presence of characters (cladists); and (ii) those that constructed relationships on the basis of both presence and absence of characters (pheneticists) (Platnick, 1989). Evolutionary systematists simply lacked any consistent and definable methodology, or fell into the pheneticists' camp.

Furthermore, the phenetic approach has been shown to be fatally flawed, largely through the work of Farris (1979, 1980, 1981, 1982a, 1983). The approach is flawed because the tree topology that is found can change according to the proportion of plesiomorphic and derived characters involved (Platnick, 1989). Figure 3.1 gives a five-taxon problem to illustrate this point. If an analysis is run on the data matrix using only the first eight characters, both distance and parsimony methods generate the same topology, [A[B[C[D+E]]]]. However, if we add a further five characters (9–13) to taxa D and E the situation changes. Under parsimony the additional five characters are seen simply as reinforcing the relationship of D+E and the topology remains as before. However, phenetic clustering changes the topology to [[[[A+B]C][D+E]]. Thus, for phenetics, the nine absences shared by A and B are seen as more important than the two presences. For this reason, parsimony analysis is now the most widely utilized method for reconstructing phylogenetic relationships.

A huge amount has been written over the past 15 years about the process of phylogenetic analysis (e.g. Eldredge & Cracraft, 1980; Wiley, 1981; Nelson & Platnick, 1981; Schoch, 1986; Ax, 1987; Sober, 1988; Maddison & Maddison, 1992; Forey et al., 1992). The present text gives only a brief account of the subject. For detailed treatment, the reader is referred to the references cited.

Cladistic analysis is the most powerful method available for reconstructing phylogeny, which it does through formal character analysis. Its assumptions are twofold: (i) there is hierarchical order to the natural

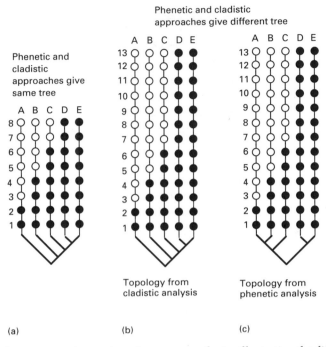

Fig. 3.1 Character matrices and resultant tree topologies illustrating the difference between cladistic and phenetic methods. Character state distributions amongst the five taxa (A–E) are illustrated by circles and are numbered 1–13. Each character occurs in two states: presence (●) or absence (○). When only the first eight characters are used (a) both phenetic and cladistic methods derive the same branching pattern of relationships. With the addition of a further five characters unique to taxa D and E (b&c) phenetic and cladistic topologies differ.

world; and (ii) this order is manifested in the distribution of characters shared amongst organisms. If both statements are true then assessment of character distributions among organisms using cladistic methodology is a valid means of establishing phylogenetic relationships. For most workers the first assumption is justified by a belief that life has originated once and that the diversity we see around us has arisen through a process of descent with modification and is thus hierarchically related. However, the specific model of how hierarchy has arisen is irrelevant in this context (Platnick, 1979; Patterson, 1980).

The simplest and most powerful approach to the analysis of character distribution is the adoption of strict methodological parsimony (Kluge, 1984). Parsimony is the approach that minimizes the number of *ad hoc* assumptions that must be made to explain a set of observations. In phylogenetic reconstruction the observations are the occurrences of characters amongst taxa and the *ad hoc* assumptions are that observed similarity of characters in two or more taxa have arisen as a result of homoplasy, not homology (for a discussion of homoplasy and homology, see below). Thus, parsimony is an analytical technique for grouping taxa that minimizes the number of instances of homoplasy that must be invoked.

Although parsimony acts to minimize homoplasy among taxa, it does not assume that homoplasies are rare in evolution, nor that evolution must have proceeded parsimoniously (Farris, 1983). It simply chooses among alternative topologies on the basis that the solution that requires the fewest additional assumptions is the best supported from available data. Parsimony does not give access to absolute truth – it only indicates 'which cladistic groupings are best supported by observations' (Sober, 1983, p. 356).

One possible criticism of parsimony is that simulation studies have shown that there are certain conditions under which parsimony analysis leads to the wrong topology (Felsenstein, 1978). Where a cladogram has short internal branches and terminal branches of extremely variable length, and character convergence occurs relatively frequently at random over the cladogram, it is possible that chance convergence (homoplasy) between long branches can mask the true phylogenetic signal. This tends to cause long terminal branches on the cladogram to pair together, irrespective of whether the taxa represent sister groups. The problem of long branches pairing is more likely to be met when dealing with molecular sequence data because of the extremely limited number of character state changes that can occur at each homologous site. Thus, homoplasy may be a severe problem for molecular data (Lanyon, 1988; Hillis, 1991; Debry, 1992). For morphological data, however, the potential number of states each character can adopt is often very much greater and homoplasy is less likely to be a problem. Furthermore, long branches can be broken up by the addition of fossil data.

Parsimony analysis stands the best chances of success if all observable data are included, unless there are good a priori reasons for believing that some classes of data may be positively misleading (Kluge, 1989). Consequently databases are often large, but the development of dedicated computer programs (see p. 45) has transformed the task into a routine process even for complex matrices.

Characters in phylogenetic analyses

All phylogenetic analyses must begin with the compilation of a data matrix based on the observed distribution of characters among taxa. Thus the first problems that are encountered are those of character definition and the recognition and establishment of homologous similarity. These are not restricted to parsimony analysis alone, but are questions that must be addressed by all methods that use morphological data.

Homology

At the core of comparative biology and systematics is the concept of homology (for an extensive review see Rieppel, 1988). In describing the morphology of an organism we break down our observations into traits or characters that are definable entities of the whole. These characters *per se* convey no information that can be used for phylogenetic reconstruction until we recognize their existence in other organisms through naming them, for it is the act of naming characters that establishes a theory of homology (Platnick, 1979). It marks the recognition that a character in one taxon represents 'the same' feature as a similar, but often non-identical, structure in another taxon.

Structures that are identical in details of form, position, and development in two or more organisms pose no problem, because all workers will agree that they represent the same entity. However, problems arise when structures have diverged in form so as to be only vaguely similar, or when different developmental pathways arrive at very similar structures. Then the identification of homology becomes much more problematic, and how we define 'sameness' becomes important.

A good example of the way in which homology hinges upon definition can be seen from the way in which character terminology has been applied to the problematic extinct group generally referred to as carpoids (Fig. 3.2). Ubaghs (1967) built up a new system of nomenclature for cornutes and mitrates, extinct groups that he thought most closely related to extant echinoderms. This nomenclatural system was useful for defining homologous elements within mitrates and cornutes, but made no hypotheses of homology that extended to other groups. Thus the single appendage was called the 'aulacophore' and the strange openings on the body were variously termed 'cothurnopores' or 'lamellate organs'. All cornutes were perceived to have aulacophores and some also had cothurnopores or lamellate organs; thus, Ubaghs was making a claim about homology within the group. However, his terminology also implied that no homologous structures could be recognized in other taxa. Jefferies (1986), on the other hand, believed that he could recognize homologies between chordates and cornutes and therefore referred to the appendage as a 'tail' and the openings as 'gill slits'. By naming the features in this way Jefferies was making a bold claim about the homology of those structures in a broader context. Thus, although the observation of characters may be relatively objective (Jefferies and Ubaghs both agree about the morphological details of these structures), the way in which different workers define and interpret characters can make a tremendous difference and is theory-laden.

Establishing the homology of characters among different taxa is a fundamental step in determining their phylogenetic relationships. But what do we mean by homology and how do we establish that charac-

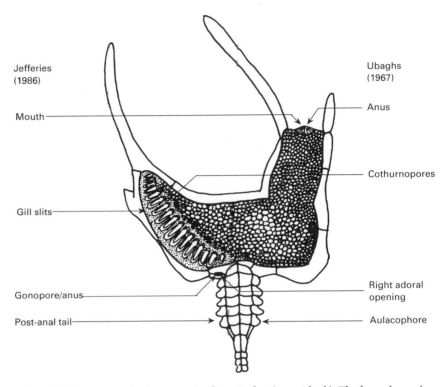

Fig. 3.2 The cornute *Cothurnocystis elizae* Bather (carpoid *s.l.*). The homology of certain structures has been interpreted differently by Jefferies (1986) and Ubaghs (1967), leading to different phylogenetic interpretations.

ters in two organisms are homologous? To an evolutionary systematist homology is defined as the same structure inherited from a common ancestor, but this leads to circularity since it is homology that allows us to postulate common ancestry. The solution seems to be to use the classic criterion of overall similarity (of form and topological position) as initial evidence for structures being homologous, and to test this using the criterion of congruence (Eldredge & Cracraft, 1980; Patterson, 1982; Rieppel, 1988; de Pinna, 1991). The congruence test equates homology with synapomorphy, so that those hypothesized homologies that are shown to have arisen once on the most parsimonious cladogram pass the test, while those that do not are assumed to be the result of convergent evolution and are referred to as homoplastic characters or simply *homoplasy*. Here the final arbiters of homology are the characters themselves, since the only test of a hypothesis of homology comes from other hypotheses of homology (Patterson, 1982; de Pinna, 1991). Consequently, the more characters used, the more stringent the test they provide of homology statements. This is important because it argues for hypotheses of homology to be evaluated on the total evidence available (Kluge, 1989).

To give an example of how this works in practice, Smith & Wright (1990) provided a cladistic analysis for 11 taxa of diadematoid echinoids, based on 25 characters (Fig. 3.3). Characters were scored as objectively as possible using classical criteria of homology. Having constructed a character–taxon matrix, it is then possible to test these hypothesized homologies by parsimony analysis. The majority of character states show a distribution that is entirely consistent with their being homologous. However, a few character states, including those of characters

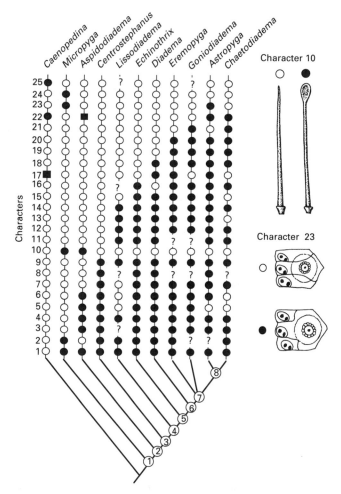

Fig. 3.3 Character matrix and cladogram for 11 taxa of diadematoid echinoids (Smith & Wright, 1990). Characters were scored as objectively as possible using classical criteria of homology. The various character states are indicated by different symbols. For example, character 10 refers to the structure of the spine tips, which are either pointed and needle-like (O) or terminate in a hoof-like structure (●); character 23 refers to the structure of the ambulacral plates, which have all three elements reaching the perradial suture (O), or small demiplates that do not reach the perradius (●). In both examples, similarity of appearance was taken as prima facie evidence for homology. (See Smith & Wright, 1990, for details of other characters.)

The cladogram was generated by a branch and bound search using the computer program PAUP (Swofford, 1993). Nodes are numbered consecutively. (See Table 3.1 for annotated list of character support for each branch of the cladogram.)

10 and 23, show disjunct distributions, indicating multiple independent derivations. Thus ambulacra with demiplating, for example, fails the test of congruence and appears to have arisen independently at least twice. It is an example of homoplasy, not homology.

Choosing and defining characters

So far I have concentrated on the methods by which phena can be recognized through morphological analysis, without properly discussing the tools that are used. Characters and character definition are fundamental to systematics, and all morphological analyses must begin by observing how characters are distributed among the members. Characters recognized in two or more taxa are statements of implied homology which are based on observed similarity but which require testing through analysis of character congruence. What precisely constitutes a character is, however, difficult to define. One general definition is that *characters are observed variations which provide diagnostic features for differentiation amongst taxa*. Characters must therefore occur as two or more states (one of the states may be 'absence'), and should be defined as objectively as possible. Characters should also be independent of one another. Different 'presence' states grouped under one character must have sufficient outward similarity to be plausibly considered as homologous features. All these conditions need to be met whether a database is assembled for phenetic or cladistic purposes.

Objectivity of definition. Different workers will perceive and define characters in different ways, because character recognition is to some extent coloured by theory and expectation (Pogue & Mickevich, 1990). Although it may seem to be stating the obvious, character states should always match observed conditions in taxa.

Characters need to be defined as objectively as possible, so as not to prejudge the outcome of the analysis. Morphologically indistinguishable features must not be coded as separate character states because of some preconceived notion of how they have evolved. For example, Landman (1989) provided a cladogram of scaphitid ammonites which included the following character: 'Umbilical diameter: a – early whorls not in contact, primitive; b,c – increasingly smaller umbilicus, progressively more derived; d – slightly larger umbilicus, reversal toward a more primitive condition'. Here the objectivity of umbilical diameter as a character is lost because Landman has unwittingly imposed his evolutionary preconception on the way it is scored. Similarly, Blake (1987) scored the same morphological state as two independent, non-overlapping characters in two taxa of asteroids because he believed a priori that the condition had arisen independently. Such prejudgement of the outcome is unjustifiable and should be avoided, because it implies that the phylogenetic relationships are already known.

Independence of characters. One of the assumptions required for parsimony analysis is that characters should be logically independent of one another. Independence for many characters is not, however, always clear-cut. The problem is simplest with molecular sequence data, where individual bases can usually be treated as independent of one another. Even here, however, there are notable exceptions, such as in the stem regions of ribosomal RNA molecules where paired sites are constrained by Watson–Crick pairing. In morphological data the problem is not as straightforward because characters forming functional complexes cannot always be considered as truly independent of one another. But this concept can be extended to the view that *all* characters are dependent on one another, since the organism itself is a single functional complex. Thus, objective boundaries cannot be drawn between integrated and independent characters (Cracraft, 1981). As a first approximation, therefore, it is probably best simply to treat all morphological characters as independent, provided that none is included within, or is logically the same as, another character.

Ill-considered character definition can also lead to compromise of the independence of character states. Combining independent conditions under one loosely defined character masks potentially useful information and may group non-homologous conditions. Conversely, oversplitting a character into many different states may result in a higher-level homology (synapomorphy) being overlooked.

Types of characters

Characters can take the form of continuous variables (ratios of length measurements, relative sizes, or shape factor products) or discrete variables (characters that occur only in a small number of states, e.g. presence/absence). They may also be coded as binary or multistate characters.

Discrete characters. Characters occurring in a small number of discrete states are the easiest to recognize and define, because their boundaries are consistently observable. In constructing a character matrix, each state is assigned a different value and taxa are scored accordingly. The most objective discrete characters are those that deal with the presence or absence of a structure. Discrete characters are more common in analyses of relatively high taxonomic rank, whereas continuous variables are much more common when dealing with phena.

Variable characters. Many characters refer to traits that show continuous variability. These deal with shape parameters, positional and relative density information, and linear ratios; they are consequently often more difficult to score objectively. One approach is simply to divide up the variable arbitrarily. Ramskold & Werdelin (1991) used this technique extensively for handling variable characters in Silurian trilobites. For example, they determined the width of the occipital ring as a percentage of the total cephalic width and then set up five arbitrary categories: 0, <32%; 1, 32–33.9%; 2, 34–35.9%; 3, 36–37.9%; 4, >38%. More often divisions are made more qualitatively. However, characters defined descriptively (e.g. narrow/wide, deep/shallow, dense/sparse) ought to be qualified and discussed so that their meaning is obvious to other workers. Fortey & Chatterton (1988, p. 172) were careful to do this in their trilobite character definitions wherever there was any doubt as to boundary divisions: 'pygidial doublure: 0, narrow; 1, wide (arbitrary definition of narrow is where the width of doublure is one-third, or less, the width of the pleural field inside doublure)'.

The problem with arbitrary divisions is that they may not be the best way to split up variable characters. Systematists can often use their experience and knowledge of character distribution to formulate a more appropriate partitioning. For example, if a variable character has a bimodal distribution within the sample taxa, creation of 10 categories of equal size is only likely to confuse matters. Conversely, if the distribution has three significant peaks, a simple median partitioning is equally inappropriate. For this reason, care is required in subdividing variable characters. The most appropriate method is to look for gaps in the unimodal distribution of the variable, which might allow objective boundaries to be drawn (Archie, 1985). Univariate plots of character distribution are therefore a useful starting point for identifying the position of gaps. More sophisticated biometric techniques can be used to analyse complex shape parameters in order to identify clusters of taxa. Michaux (1989) used principal component analysis to identify size- and shape-related clusters of taxa, scoring each cluster as a different character state. Canonical scores derived from a discriminant analysis have been used directly as character states by Cheetham (1987) and Budd & Coates (1992), but these are neither truly independent of one another nor do they deal with putative statements of homology. They are thus inappropriate for cladistic analysis.

Binary and multistate characters. Characters with only two alternative states are termed binary characters; those with three or more are multistate characters. To a large extent, characters define themselves – whether a character should be coded as binary or multistate is usually obvious.

There are three alternative approaches to multistate data:
a The states can be organized into a single ordered character with a predetermined sequence of steps on some a priori evidence (e.g. ontogeny or stratigraphic distribution).
b The states can be scored separately as a set of presence–absence binary characters.
c The states can be scored as a single multistate character in which any character state is free to transform into any other.

The important difference is that under option **a** a change from state 1 to state 3 along a branch in the cladogram counts as two steps whereas under **c** it counts as only a single step (Fig. 3.4). Whether characters are ordered or not can therefore affect the outcome of the parsimony analysis.

It was believed that ordering multistate characters before analysis helps to improve cladistic resolution (Mickevich, 1982). However, Hauser & Presch (1991) found no correlation between tree resolution and whether multistate characters were treated as ordered or unordered;

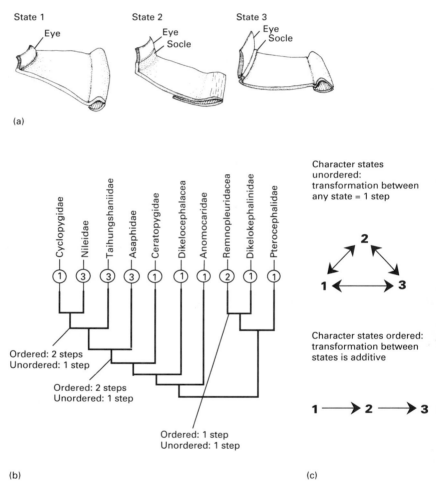

Fig. 3.4 Multistate characters and their coding. The structure of the eye socle in asaphine trilobites was coded as a multistate character by Fortey & Chatterton (1988) and the three possible character states are illustrated. The distribution of these three character states is indicated on the cladogram, which represents one of four equally parsimonious solutions (see Chapter 6 and Fig. 6.8). Each branch where the character state changes is indicated on the cladogram, and the number of steps that must be assumed when the character is treated as either unordered or ordered are indicated.

nor did they find an increase in the number of equally parsimonious trees when data were unordered. It would therefore seem better to treat all multistate characters as unordered and let the parsimony analysis derive the phylogenetic pattern of that character, as Hauser (1992) has argued. The only danger is that this approach favours character state trees that maximize consistency in preference to observed similarity (Lipscomb, 1992).

Should characters be weighted a priori?

Having scored the characters and constructed the matrix, the next question is whether or not to weight characters prior to analysis. In fact, all analyses apply weights to characters, but many cladists apply equal weight to all characters irrespective of any prior assumptions. Characters 'weight themselves, by associations that are beyond the bounds of chance' (Patterson, 1982). This is certainly the most conservative approach, but a data matrix that includes a mixture of highly homoplastic characters along with some fully determinate characters may generate multiple equally parsimonious solutions in an equal weights analysis, because high levels of homoplasy tend to swamp any phylogenetic signal.

Some systematists have thought it appropriate to apply a weighting scheme prior to analysis to help reduce problems of homoplasy. For example, Fortey & Chatterton (1988) employed a form of weighted analysis in their study of the trilobite suborder Asaphina, by omitting characters that they believed might be more prone to homoplasy. They carried out two analyses: one based on all available characters, the other using a subset of the characters that they believed were of particular importance phylogenetically. In this case the two approaches resulted in only small differences in topology.

The problem with weighting a priori is that, in the absence of a cladistic analysis, there is no means of judging which characters should be preferentially weighted and by how much. The experienced systematist with first-hand knowledge of thousands of taxa may be in a position to judge which are good characters and which are highly variable from knowledge of their distribution, but this may be difficult to justify to others and implies that the phylogeny is already known. Therefore a priori weighting is not appropriate and, if weighting is to be applied, it should be done a posteriori (see p. 53).

Missing data

In some cases, it will prove impossible to score certain characters for all taxa. This can arise for three reasons:
1 The character is polymorphic in the taxon.
2 The character state is unknown because of incomplete preservation.
3 The character state is inapplicable to the taxon.

Each represents a different type of ambiguity and should be treated differently.

Polymorphism. A character is polymorphic in a taxon when different phena assigned to that taxon display different states of the same character. Either the taxon is monophyletic and character variability has arisen through divergent evolution within subgroups, or the taxon is polyphyletic and phena with different character states have been grouped in error (Nixon & Davis, 1991) (for definitions of monophyly and polyphyly, see p. 74). Only additional cladistic analysis of the taxon will determine which is the correct explanation.

If the taxon is monophyletic and cladistic relationships of phena within the taxon are known, then characters observed in the most basal phena should be taken as general for the taxon as a whole. Alternatively, polymorphic taxa could be broken up into their constituent elements, each of which is uniform in its coding. If no cladistic hypothesis is available, other alternatives must be sought. For example, the type species alone could be used to determine character states since, in nomenclatural terms, they are the objective name bearers for higher taxa.

Fortey & Chatterton (1988) used stratigraphic information and a more intuitive approach to determine which of several alternative character states within a single higher taxon of trilobites should be taken as representative of the taxon as a whole. Asaphacid trilobites, for example, are predominantly devoid of tuberculate sculpture. There is, however, one genus, *Norasaphus*, which appears stratigraphically late in the group's history and which has pronounced tuberculate sculpture. In the absence of any cladistic hypothesis of relationships within asaphacid trilobites, Fortey & Chatterton treated *Norasaphus* as a derived form and scored asaphacids as non-tuberculate. This of course assumes that character states appearing earlier in the fossil record are more primitive. Such claims remain to be tested by parsimony analysis but, solely as an interim hypothesis, are reasonable.

Incomplete information. Missing data can be a significant problem for fossil phena. Often, crucial structures are missing or inadequately preserved in fossils, such that they cannot be scored. The inclusion of many unknown character states in a data matrix can significantly increase the number of equally parsimonious trees generated from the data (Rowe, 1988). This in turn results in consensus trees that contain more multiple branching points (polytomies) and consequently less hierarchical information. However, even poorly known taxa may be crucial for the results of a phylogenetic analysis (p. 64), and should not be omitted on a first pass through the data. In practice, then, a first analysis should be run that includes all relevant taxa, irrespective of how incompletely known they are. If it is found that one or more of these taxa are topologically unstable and are consequently collapsing an otherwise well resolved cladogram to a polytomy, then they can justifiably be excluded from further analysis as uninformative.

Inapplicable characters. Some complex structures can be scored for several highly informative characters. Yet if a taxon lacks this structure (through secondary loss or primary absence) it is logically impossible to score those characters. For example, the traits associated with crinoid pinnule structure cannot be scored for crinoids that do not possess pinnulate arms. Similarly, the presence or absence of teeth can be recorded objectively for mammals and birds, but specific dental characters that are useful for differentiating major groups cannot be scored in certain groups of mammals or in extant birds because teeth have been lost.

Inapplicable characters cause problems because they cannot be scored as any of the optional character states (Platnick *et al.*, 1991). Treating them as unknown is not appropriate, since computer programs will assign one of the alternative states to each 'unknown' in the data matrix and this can lead to spurious resolution and relationships. Where character states are inapplicable to certain taxa, it is often possible to recast character definitions to avoid this problem. This can often be achieved by amalgamating two or more binary characters into a single multistate character. Thus, for example, instead of scoring for two characters – 'pinnules absent/pinnules present' and 'pinnules ending in a hook/pinnules ending bluntly' (with those taxa lacking pinnules scored as '?') – the character can be reformulated as a single three-state character – 'pinnules absent; pinnules present and terminating in a hook; pinnules present and terminating bluntly'. With care, most situations in which character states are logically inapplicable can be avoided.

Presentation

Having itemized as many different characters as possible, each taxon is then scored accordingly. In this way a comprehensive taxon–character matrix can be constructed. It is important that these primary data are presented in a way that others can assess and use. Unless one is a specialist totally conversant with the taxa under study, it can be very difficult to judge the strength of signal residing in the data. Other workers may want to check the results or use different approaches, or add or revise character scores. Therefore the matrix should always be presented either alongside the cladogram, or as an integral part of the cladogram, as in Fig. 3.3. Simply listing the characters supporting each node in the cladogram is not adequate because the primary data matrix cannot be reconstructed if there is homoplasy or missing information. More importantly, it is essential that other workers are made aware of any inconsistency in characters listed as supporting a node. For example, a character listed as a synapomorphy for a node may be present in all terminal members of the clade (and thus highly consistent) or may be prone to considerable homoplasy, such that few of the terminal members retain it. In the latter case, a false aura of reliability is imparted.

In addition, it is a great help if the character changes supporting each branch are itemized. Rowe (1988) proposed a set of conventions that help to standardize how such data are presented. Thus, each node should be numbered on the cladogram and characters listed and numbered. A table can then be presented listing the character state changes that support each node. Table 3.1 gives an example of a character matrix and listing presented in this way for the cladogram and data matrix in Fig. 3.3. Other examples are given by Tassy (1991), Forey (1991), Cloutier (1991) and Ramskold & Werdelin (1991).

Table 3.1 Characters identified as supporting the cladogram shown in Fig. 3.3. All characters were treated as unordered and the tree rooted on *Caenopedina*. For changes affecting multistate characters, the derived state at that node is given in parentheses after the character number. Reversals of characters that have already appeared lower in the cladogram are indicated by a minus sign. Character changes of a hierarchical position that is uncertain – due to scoring of the character as unknown in one or more adjacent sister groups – are marked with an asterisk* and listed at the node where they first appear. However, since the change may have occurred earlier in the cladogram but cannot be proved because of missing data, the earliest possible node at which such a change could have occurred is also given in brackets

Node 1: 1, 2
Node 2: 3, 4, 5, 6
Node 3: 7, 8, 9
Node 4: 11, 12, 13, 14
Node 5: 15, 16*(node 4)
Node 6: 17(1), 18
Node 7: 19, 20
Node 8: −5, 22(1)
Caenopedina: 10, 17(2), 22(1), 25
Micropyga: 10, 23, 24
Aspidodiadema: −2, 10, 22(2)
Centrostephanus
Lissodiadema: −5, −6, −7
Echinothrix
Diadema: −16
Eremopyga
Goniodiadema: −17(0), 21
Astropyga: 23
Chaetodiadema: −5, −6, −13, −15, −17(0), 21

Cladograms and their construction

Numerical methods

The compilation of character data for the suite of taxa under study marks the first stage of a cladistic analysis. The next stage is to turn this into a cladogram. It is important to use all available information, since the greater the number of characters, the more stringent is their test of homology (see p. 63). For example, Lauterbach (1980) made the case that trilobites were a paraphyletic grouping, by identifying three characters that linked olenellid trilobites to limulids. He used this evidence to argue that olenellids were the sister group to the Euchelicerata plus Pantopoda (Fig. 3.5), a suggestion accepted by Ax (1987). The three characters uniting olenellids and limulids all relate to the development of a thoracic spinous process in the two groups. However, Fortey & Whittington (1989; see also Fortey, 1990b, and Fortey & Owens, 1991) compiled a much more complete data matrix for the taxa in question and were able to show that olenellids shared many more derived characters with other trilobites than they did with limulids (Fig. 3.5). The addition of more data revealed that the similarities in thoracic structure between olenellids and limulids were the result of homoplasy. Cladistic hypotheses stand the best chance of being correct if they are formulated on the basis of all available data.

Although small data matrices can be analysed by hand, any reasonable-sized data matrix requires a parsimony computer program to ensure that the topology best supported by the data is found. There are currently four widely used computer programs for phylogenetic analysis using parsimony:

1 *PHYLIP* (*phy*logenetic *i*nference *p*ackage). Available from J. Felsenstein, Department of Genetics SK-50, University of Washington, Seattle, WA 98195, USA. This is compiled for IBM-PC and Macintosh microcomputers and contains a variety of phylogenetic programs, including bootstrapping facilities.

2 *PAUP* (*p*hylogenetic *a*nalysis *u*sing *p*arsimony). Written by D.L. Swofford (1985, 1993) and distributed by Illinois Natural History Survey, 607 E. Peabody Drive, Champaign, Illinois 61820, USA. There are two versions: an earlier IBM-PC version (1985), and a much more extensive version for the Apple Macintosh (1993) which includes bootstrapping and tree asymmetry options. Both versions include simple options for character and taxon editing.

3 *HENNIG86*. Obtainable from J.S. Farris, 41 Admiral Street, Port Jefferson Station, New York, NY 11776, USA. Compiled for IBM-PC computers only and including options for tree manipulation and character editing.

4 *MacCLADE*: Analysis of phylogeny and character evolution, version 3.0 (Maddison & Maddison, 1992). Available from Sinauer Associates, Sunderland, Massachusetts 01375, USA. This is not really

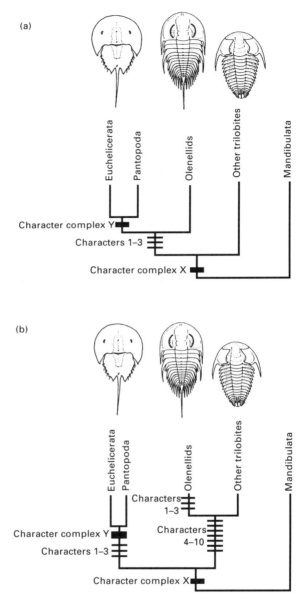

Fig. 3.5 The importance of total evidence. Two competing hypotheses for the relationships of olenellids. (a) Lauterbach's (1980) proposal has trilobites as paraphyletic with the *Olenellus* clade as sister group to the euchelicerates and pantopoda. Three morphological characters were cited in support of this group. (b) The relationships found when many more morphological characters are considered: trilobites are a well-supported monophyletic group. (From Fortey (1990b), where the supporting characters are listed.)

a powerful tree-finding program but it does provide a very useful series of options for exploring suboptimal trees and character distributions. It is fully interchangeable in format with PAUP.

Hand-crafted cladograms (where characters supporting nodes are identified intuitively) justify, but do not test, phylogenetic hypotheses. They identify apomorphies that support a particular branching pattern and thus demonstrate the grounds on which the cladogram was constructed. However, if the data matrix is large and there is homoplasy, then an intuitive approach may not necessarily identify the most parsimonious solution. For this reason, numerical computer techniques are to be preferred.

It is worth stressing that the cladograms that result from this process are only as good as the data fed into them. It is highly recommended that the characters are carefully re-examined in the light of the resultant cladogram. Plotting the distribution of individual characters on the cladogram, and noting where homoplasy is implied, gives insight as to where there might be problems of character definition. Similarly there may be nodes that are poorly resolved and which prompt a re-examination of the taxa for additional characters that might shed light on the problem. If poor resolution is due to conflict in characters, these characters may merit reconsideration.

This does not of course mean that one is entitled to invent or otherwise manipulate character states by, for example, coding the same trait as two independent characters in different parts of the cladogram, or coding absences in some taxa as secondary reversals. The data matrix must be a true summary of observed character distribution. Nevertheless, some character definitions are open to more than one interpretation and carrying out several runs using alternative codings can often be helpful.

Finally, available characters for many groups may simply not provide satisfactory resolution (e.g. Pandolfi, 1989). However, when such poor resolution arises because of conflict between character states, this itself is revealing something about character evolution in the group, even if it is preventing fine-level resolution of phylogenetic relationships.

Evaluating the information content of cladograms

In any real data matrix of any size there is usually a substantial amount of homoplasy which has arisen through convergent evolution or reversal. From first principles some forms of character change are more likely to occur than others. For example, the loss of a complex character is more likely than the independent evolution of that character. However, homoplasy can in practice only be demonstrated through parsimony analysis; thus there are no grounds for prejudging the outcome by a priori down-weighting of certain characters. It is often helpful to have some measure of how much homoplasy exists in a data matrix or, to put it another way, how much phylogenetic information there is. Computer algorithms will construct cladograms from random data and so it is important to have a means of measuring how much phylogenetic signal resides in the data. Several approaches can be used.

The consistency and retention indices. Two common measures of homoplasy can be calculated from the distribution of character states on a cladogram: the consistency index and the retention index. The consistency index (CI) is calculated from the number of homoplasies that must be assumed in the most parsimonious solution. In the simplest case, if a character has two states and the most parsimonious tree requires that that character changes only once (i.e. the derived character is found within every member of a monophyletic clade), the consistency index for that character is then 1.0. If, however, the most parsimonious solution indicates that that character has changed twice in the tree (either through secondary loss of the derived state within the clade or convergent evolution in a second clade), then the character has a CI of 0.5. The consistency index for the cladogram as a whole is equal to the total number of derived character states scored in the matrix divided by the number of steps required to produce the tree. It therefore decreases as the level of homoplasy increases.

The retention index (RI) measures the proportion of terminal taxa that retain the character identified as a synapomorphy for that group. Thus, if a character identified as a synapomorphy for a clade is present in all the terminal taxa, it is given an RI of 1.0, whereas if the character, through later transformation or reversal within the clade, is present in only 50% of the terminal taxa, its RI is 0.5. More detailed discussion of the derivation and use of these indices is given by Farris (1989), Swofford (1991), and Forey *et al.* (1992).

Both measures are affected by the inclusion of autapomorphies (characters unique to just one terminal taxon). Therefore if topologies are to be compared, it is worth removing autapomorphies before calculating either the CI or RI. Also, the amount of homoplasy, as measured by CI, generally increases as the number of taxa included increases (Sanderson & Donoghue, 1989) (Fig. 3.6). As data matrices become larger and the taxa more densely sampled, so it becomes easier to distinguish homoplasy from homology.

Testing the support for individual branches

Parsimony programs can generate cladograms even where there is very little hierarchical structure of characters in the data matrix, therefore it is important to identify which parts of the tree are well supported and which are weakly supported. Here there are also several approaches (reviewed by Swofford & Olsen, 1990), most of which have been developed for molecular data in the first place.

The first step is always to check the synapomorphies assigned to each branch. When cladograms are computer generated it is very important that the individual character states at each node are checked by hand, because the computer program will determine at what level character state changes are placed on the cladogram according to a preselected optimization procedure. Changes may either be forced to

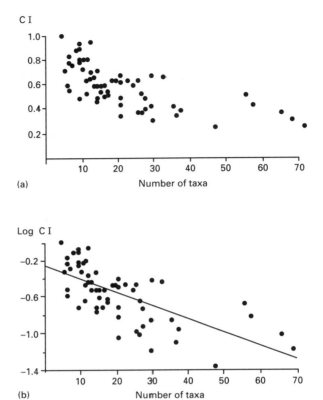

Fig. 3.6 Consistency index as a correlate of the number of taxa included in a cladistic analysis (Sanderson & Donoghue, 1989). The consistency index (C I) of the most parsimonious cladogram is plotted against the number of taxa included in the analysis for 60 data sets. (a) Raw data. (b) Log transformed consistency index.

the earliest possible node, or delayed to the latest possible node (Fig. 3.7). If there are data missing from the matrix, some groups may end up being supported only by an inferred character state change for which there is no real evidence.

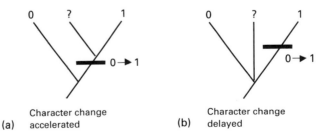

Fig. 3.7 Where a character state change cannot be unambiguously tied to one node in the cladogram, because of missing data, it will be positioned by convention according to the optimization criterion in effect. This can result in groups being supported solely on the basis of inferred changes (a).

Jack-knifing. This method consists of breaking up the data matrix into subsamples, repeating the parsimony analysis for each subset, and then seeing how much the structure of the tree is altered. This can be done by using the full suite of characters and subsets of taxa, or by using the full suite of taxa and only a subset of characters. However, by reducing the size of the data matrix the chances of obtaining the original tree are greatly reduced and the latter option is probably too stringent a test. If a topology is resistant to this treatment, there is certainly a strong signal in the data. An alternative approach involves removing one taxon at a time and calculating the overall variance associated with the slightly reduced data matrix. This provides a statistical test of how much signal is present in any data set (Lanyon, 1987; Swofford & Olsen, 1990).

Bootstrapping. With this method, characters are taken at random from the data matrix to construct a new data matrix of the same dimensions as the original. The new data matrix may include the same character more than once, while others may be excluded altogether, because characters are chosen at random. This new matrix is used to calculate a topology and the process is repeated many times. The number of times a particular branch appears is used as a measure of how strongly supported that branch is. Thus, a branch appearing in all 1000 bootstrap replicate runs is 100% supported, whereas one appearing in only 200 is 20% supported and therefore less reliable. The computer parsimony program PAUP version 3.1 (Swofford, 1993) has a simple option that allows bootstrap support to be calculated for cladograms.

The relationship between bootstrap per cent support (P) and the probability that a particular clade is present in the true tree (assuming that the character matrix is representative) is not, however, straightforward. The index P is affected by many parameters, including the size of the data matrix, the number of iterations performed, and the tree topology and position of the group of interest within the tree (Hillis & Bull, 1993). Hillis & Bull found that, under typical conditions encountered by systematists, bootstrap percentage supports greater than 50% are consistently lower than the probability that the corresponding branch is correct. Bootstrap results should therefore be interpreted only as highly conservative estimates of the accuracy of tree structure, and the values cannot be directly compared among studies where parameters differ. To counter some of these problems Felsenstein & Kishino (1993) have recommended that, instead of using P as a measure of the probability that the data has correctly identified a group in the true tree, the inverse $(1-P)$ should be used as a conservative assessment of the probability of getting that much evidence favouring a group if it is not present.

Suboptimal tree distributions. Both jack-knifing and bootstrapping are resampling techniques that provide a heuristic measure for the strength of support for individual branches. A different approach is to see how many additional steps are required before the most parsimonious topology starts to break down. If the data matrix is sufficiently small (i.e. 12 or fewer taxa), an exhaustive search of every possible topology can be carried out within a reasonable time using a computer program. All trees within a certain number of steps of the most parsimonious can be saved and their topologies examined to see how many additional *ad hoc* assumptions about homoplasy need to be invoked to derive an alternative (competing) hypothesis. Novacek (1992a) evaluated the robustness of his cladogram of mammal relationships in this way (Fig. 3.8). He found that the group Paenungulata

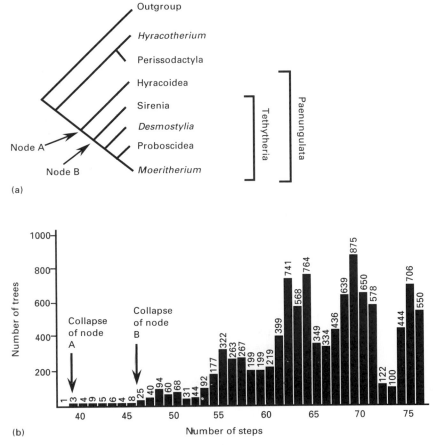

Fig. 3.8 The use of suboptimal trees to assess the support in the data matrix for individual branches (Novacek, 1992a). (a) The most parsimonious cladogram derived for selected Recent and fossil mammals (see Novacek, 1992a for full details). The two nodes that define the groups Paenungulata (node A) and Tethytheria (node B) are indicated. (b) The distribution of all cladistic trees derived from an exhaustive search of the data matrix. A single, most parsimonious tree is found at 38 steps. However, three trees one step longer fail to support the Paenungulata grouping at node A. The tethytherian clade appears in all trees up to 45 steps in length. Node B is very much more strongly supported than node A.

(Hyracoidea, Sirenia, and Proboscidea) was very weakly supported and broke down with just one additional step. By contrast, the group Tethytheria did not collapse until a further eight steps were added to the most parsimonious tree, indicating that this grouping is much more strongly supported in his data.

Tree-length distribution. Sometimes, when very many equally parsimonious solutions arise from a data matrix, it is useful to know whether there is any hierarchical pattern to the character distribution at all, and at what level in the cladogram it resides. Huelsenbeck (1991a) and Hillis (1991) (see also Hillis & Huelsenbeck, 1992) have developed a method for using tree-length distribution as a measure of how much phylogenetic signal resides in a data matrix. The basic idea is simple but requires that all possible trees be calculated for a set of taxa. At present this is an option found only in the computer program PAUP version 3 (Swofford, 1993) and is only feasible for 12 taxa or fewer. The method requires that the number of steps (character changes) required for every possible topology is calculated. These are then used to construct a histogram of tree-length distribution.

Random data generate a tree-length distribution with a normal distribution, whereas data with a strong phylogenetic signal have a tree-length distribution that is strongly skewed to the left (Fig. 3.9). The g1 statistic, calculated by PAUP, gives a measure of this skewness. The tree-length distribution measure is independent of tree topology (Hillis, 1991). However, only a small amount of phylogenetically informative data (as little as 20%) need be included in an otherwise random data matrix for the tree-length distribution to be left-skewed. For this reason, Hillis (1991) suggested adopting the following procedure:

1 Run an exhaustive search of all possible trees and construct a tree-length distribution histogram. If it shows an effectively normal distribution, stop – there is no phylogenetic signal; otherwise proceed to **2**.
2 Remove the best supported pair of taxa from the cladogram and repeat **1**.

This approach identifies those taxa for which there is reasonable support from the data and to recognize which taxa cannot be resolved on available data.

Multiple equally parsimonious solutions

In phylogenetic analyses the data often allow full resolution of relationships for only part of the cladogram. More than one equally parsimonious solution may be found, either because of a lack of informative characters or because of character conflict. These can be considered as rival trees that explain the character distribution equally well. There are several options that can be followed under such circumstances, as follows.

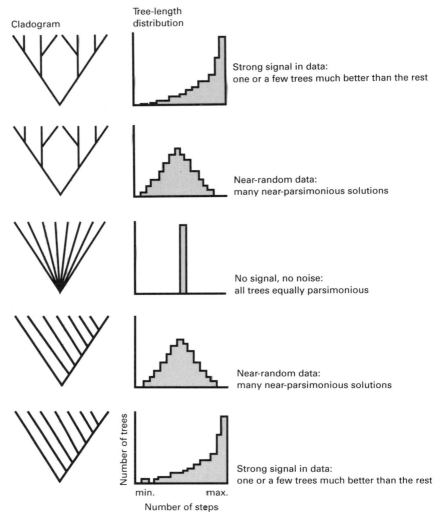

Fig. 3.9 Tree-length skewness as a measure of phylogenetic signal (Hillis, 1991). The model behaviour of tree-length distribution is expected to vary according to the amount of phylogenetic signal in the data. A strong phylogenetic signal generates a left-skewed tree-length distribution independent of cladogram structure.

A posteriori weighting. If alternative topologies arise from character conflict (i.e. different characters support different and incompatible groupings), then one option is simply to select one of the trees. This may be done on the basis of an assessment of the relative weights (reliability) of the characters in conflict. Given two characters in conflict – one a trait with high structural complexity and the other either very simple or representing a secondary loss – the former can be considered less likely to have evolved in parallel and to represent a more reliable apomorphy. Thus, de Pinna (1991) argued that wherever there was ambiguity of character distributions, the solution that favoured reversal of a structure by loss over parallel evolution should be preferred. Provided the alternative trees are given and the reasons for selecting one are clearly stated, a posteriori selection is valid.

A more objective method for applying character weights is to base the weighting scheme on the consistency or retention index derived from a first (unweighted) pass through the data. A character that shows no homoplasy on the cladogram is given maximum weight, whereas characters that must have evolved more than once on the cladogram are down-weighted according to the number of independent origi-nations that they are identified as having. In this way characters that are more prone to homoplasy are given less weight. The parsimony analysis is then repeated with the characters assigned their new weights either once, or until no change in topology results. This method is referred to as the *successive weighting method* of Farris (1969) (see also Carpenter, 1988).

Although this is an objective method of weighting it is not without problems. If there are many missing data, some characters may be artificially up-weighted simply because they are missing in a proportion of the taxa, thus biasing the outcome (Novacek & Wheeler, 1992). Also, there is the possibility that the outcome of successive weighting will depend heavily upon the starting (unweighted) topology. If, indeed, homoplasy is extensive and the starting topology is misled by homo-plasy, then successive weighting may simply converge on a wrong topology.

Consensus trees. In cases where there are very many equally par-simonious trees it is helpful to construct a consensus tree that sum-marizes the parts in all the rival cladograms that are in agreement.

Swofford, 1991 gives a very clear account of consensus methods and their relative merits.

The simplest option is to construct a *strict consensus tree*, which identifies only those nodes that are common to all rival cladograms. However, a strict consensus can be too strict in that the presence of only one or two taxa whose position is unstable may be sufficient to collapse the structure into an unresolved bush even if the other taxa are consistently related in all cladograms (Adams, 1986; Funk & Brooks, 1990).

An *Adams consensus tree* often preserves more of the original struc-ture of rival cladograms than does a strict consensus tree. Any taxonomic statements shared by the cladograms being compared are included, regardless of whether they constitute completely uncon-tradicted components (Adams, 1972; Mickevich & Platnick, 1989). The drawback of Adams consensus trees is that they may identify groups that are not found in any of the cladograms that they are constructed from.

Finally there is the *combinable-component consensus tree* of Bremer (1990), which is like a strict consensus tree but allows components that are compatible, either because they have no taxa in common and thus do not conflict, or because one group is a subset (more resolved) of the other. This method provides more resolution than a strict consensus tree and, in contrast to the Adams consensus tree, does not generate clusters that are absent from the rival cladograms from which it is constructed.

The most important point to remember about consensus techniques is that they should not be interpreted as, or used in constructing, phylogenies – they are simply 'statements about areas of agreement amongst trees' (Swofford, 1991, p. 311). Thus, a polytomy on a consensus tree represents only ignorance of relationship, not simultaneous branching. Therefore only fully resolved portions of a consensus tree should be calibrated against the fossil record and used as the basis for constructing an evolutionary tree.

Majority rule trees. Instead of trying to identify elements common to all rival cladograms, an alternative is to choose the topologies that appear most often among the alternative cladograms. Thus, only groups that appear in more than some specified percentage of all rival cladograms are used to construct a majority rule tree (Margush & McMorris, 1981). Typically the cut-off point is 50%, so that any group that appears in the majority rule tree is found in more than half of the competing cladograms.

The drawback here is that the decision on which topology is chosen is not made on the relative strengths or weaknesses of the characters in conflict, but by more or less arbitrary convention.

Additional evidence. Given two or more rival cladograms, the one that offers the most congruence with stratigraphic occurrence data, or biogeographic data, is the better supported, since this requires the fewest additional *ad hoc* assumptions about range extensions or missing geographic data (see Chapter 6).

In conclusion, consensus trees are unsuitable as a basis for constructing evolutionary trees because they represent only partial phylogenetic hypotheses. Majority rule trees are also a poor option because they offer little more than convention to select from among rival cladograms and take no account of relative character strengths. Thus, evolutionary trees should be constructed from cladograms that have been selected from among rival solutions, either through a posteriori assessment of characters, or through additional, non-morphological data (where available) that favour one solution over the others.

Rooting and character polarization

Parsimony analysis programs cluster taxa according to their characters so as to minimize the total number of character changes. In doing so they generate unrooted cladograms, which tell us about taxon similarity in the broad sense and also allow us to discriminate homology from homoplasy. However, until the cladogram can be rooted, the morphological information it carries cannot be exploited to the full.

Rooting is the process by which we distinguish derived (apomorphic) from primitive (plesiomorphic) states through the establishment of character polarity. Rooting is important because: (i) it establishes the basis for distinguishing monophyletic groups from other kinds of groupings; and (ii) it allows the historical evolution of character states in a phylogeny to be deduced.

A great deal has been written about methods for determining character polarity, which has been perceived as 'one of the major problems for the phylogeneticist' (Schoch, 1986, p. 134; see also Temple, 1992). Methods that have been suggested include: (i) commonality (e.g. Crisci & Stuessy, 1980); (ii) outgroup comparison (Watrous & Wheeler, 1981; Arnold, 1981; Farris, 1982b; Maddison *et al.*, 1984; Clark & Curran, 1986); (iii) ontogeny (Nelson, 1978b; Nelson & Platnick, 1981; Patterson, 1982, 1983; Wheeler, 1990); and (iv) stratigraphic sequence information (Szalay, 1977; Nelson & Platnick, 1981; Schoch, 1986; Fortey & Chatterton, 1988; Fortey, 1990a).

There is little doubt that the technique with the greatest power and fewest a priori assumptions is the outgroup method (Farris, 1982b; Clark & Curran, 1986). This works by designating two groups: an *ingroup* of taxa whose relationships are under investigation, and an *outgroup* of taxa whose role is to establish what state each character takes at more general levels in the hierarchy (Fig. 3.10). The outgroup comprises taxa that are sufficiently closely related to the ingroup such that character homologies can be established without difficulty. When a character occurs as two states in the ingroup, but only one of these is found in the outgroup, the assumption is that the character found only in the ingroup is derived with respect to the more generally distributed state By taking all characters into consideration the best supported root for the ingroup is found by identifying the branch in the ingroup with the fewest derived character states.

The simplest and most effective method of outgroup rooting is a simultaneous and unconstrained parsimony analysis of ingroup and outgroup taxa together. In this way the only assumption is that the root is basal to the ingroup. No prior assessment of character polarity is needed, because all characters are treated as unordered, and the method provides the greatest likelihood of finding the globally most-parsimonious arrangement. The resulting unrooted cladogram can be polarized by treating the outgroup as more basal to the ingroup. There is therefore no need to determine character polarity beforehand – their polarity is set when the unrooted network is rooted on the outgroup.

Even though the method assumes some prior knowledge of outgroup and ingroup relationships (namely, that appropriate outgroup taxa to the ingroup have already been identified), this does not prejudge the outcome. For example, what if a mistake is made and the ingroup is

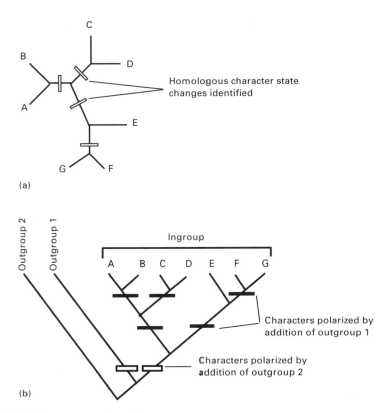

Fig. 3.10 Outgroup rooting. (a) An unrooted cladogram for taxa (A–G) allows homologous character states to be distinguished from homoplastic similarity but does not indicate the evolutionary direction of transformation between character states. (b) Rooting using a closely related outgroup (outgroup 1) allows character state transformations to be polarized within the ingroup. The addition of a second outgroup (outgroup 2) polarizes characters in the basal branch and identifies homoplasy between outgroup 1 and the two basal branches of the ingroup.

non-monophyletic, such that one of the outgroup taxa falls within the ingroup? As global parsimony is used and no constraints are imposed that would force the outgroup to stay together, this will pose no problems. The most parsimonious solution will show that the initial assumption was wrong, by placing the rogue outgroup taxon within the ingroup. If enough outgroups are included, the newly expanded ingroup will still form a clade on the unrooted tree and can be rooted by reference to the remaining outgroup taxa.

Choice of outgroup taxa can be important for the success of this method. Outgroup polarization works best if more than one outgroup taxon is used and if the outgroups are the most closely related taxa to the ingroup (Maddison *et al.*, 1984). Having two outgroup taxa allows synapomorphies of the ingroup to be distinguished from apomorphies of its closest outgroup. The closest taxon will sort out polarities for most characters, leaving only those in the basal segment unpolarized. The next closest taxon polarizes these basal characters and identifies any homoplasy that might exist between the first outgroup and the ingroup members (Fig. 3.10). The closest sister group to the ingroup will have the greatest number of characters that can be scored. For more distantly related taxa, divergent evolution will tend to reduce the number of identifiable homologous character states that can be usefully compared. Thus, for an ingroup composed of placental mammals, more characters could be usefully compared with a monotreme than with a fish or lizard.

Outgroups based on a single taxon, or taxa that are too distantly related, can produce spurious rootings (Maddison *et al.*, 1984) and should be avoided. In some circumstances, however, it may be appropriate to construct a single hypothetical outgroup whose character codings represent the presumed plesiomorphic condition deduced from a number of outgroup taxa.

The strength of outgroup method is that characters can be left unpolarized in the parsimony analysis and the tree rooted a posteriori by designating one or more taxa as outgroup. All other methods of rooting require character polarity to be set beforehand.

Fossils and cladogram rooting

In the past, some workers (e.g. Harper, 1976; Gingerich & Schoeninger, 1977; Szalay, 1977) have argued that the fossil record provides a method for determining character polarity and thus rooting cladograms: 'for groups with an extensive fossil record the character state first appearing in the record is likely [to be] the ancestral state' (Harper, 1976, p. 185). A similar theme has been developed by Fortey & Chatterton (1988, p. 166) who 'use stratigraphical criteria for the identification of primitive character states within particular families' among the trilobites they studied. Fortey & Chatterton were interested in the phylogenetic relationships of 'family' level taxa but found that some character states were variable among the members included within one family. They made two *ad hoc* assumptions: (i) that the family taxa they were using as terminal units of analysis were monophyletic; and (ii) that the oldest members in each family showed the primitive characteristics of the family. As a first approximation this may be appropriate, but a cladistic analysis of phena in each family is required to test whether these assumptions are valid.

The use of fossil data for determining character polarity has been extensively criticized by Eldredge & Cracraft (1980) and Nelson & Platnick (1981) for its simplistic assumption that early equals primitive. As Nelson & Platnick (1981, p. 333) point out, evolutionary sequences of ancestor–descendant cannot simply be read from the rocks – fossils first have to be ordered into some form of hierarchy on the basis of their morphology. The phylogenetic relationships of fossils must be understood before their stratigraphic record can be interpreted and used to map the distribution of character states through time. No one would claim that fossils always appear in their correct stratigraphic order; therefore each case must be argued on its own merits. The fossil record therefore provides data on polarity only on an *ad hoc* basis. Furthermore, Nelson & Platnick (1981) noted that any discrepancy between the order in which taxa appear in the fossil record and their order of appearance as determined from cladistic analysis of the morphological data, can always be attributed to problems of the fossil record. Mismatch may simply imply that the fossil record is not as complete as was expected. Finally, even if fossil occurrence data provide a guide to character polarity that is more often correct than not, we still need some criterion to judge when we should trust the evidence of the fossil record and when we should reject it.

As the most effective parsimony methods no longer require character polarity to be determined prior to cladogram construction, the whole question of whether fossil data can provide the polarity for characters becomes largely academic. It is now possible to test, with empirical data, how well the fossil record has captured the order of character acquisition through time (see p. 123).

When it comes to understanding why palaeontological data should not be ignored when rooting cladograms, stratigraphic occurrence *per se* is a complete red herring. It is not the stratigraphic age of a fossil that is important, but the position it holds on a cladogram and the information value of the character assemblages that it preserves. It is simply that many extant groups diverged from their living sister taxon far back in geological time, and consequently often show highly divergent morphologies. Rooting such an ingroup using an outgroup composed of modern taxa alone may mislead because of homoplasy that has arisen since ingroup and outgroup diverged. Furthermore, the number of homologous characters that can be scored in the outgroup is likely to decrease with increasing divergence. Therefore, using taxa that lie closer to the basal node (fossils) reduces the chances of mistaking homoplasy for synapomorphy and improves the likelihood of arriving at the correct topology (Huelsenbeck, 1991b; Wilson, 1992). The inclusion of fossil taxa for both ingroup and outgroup in an unconstrained parsimony analysis is therefore highly recommended.

Fossils and phylogenetic reconstruction

The failure of palaeontological methods in phylogeny reconstruction

Fossils, as the only direct evidence we have of past life-forms, have an important role in historical geology, for example in dating and correlating rocks, and reconstructing past environments. It was also believed that they were crucial for the correct reconstruction of phylogenies (e.g. Simpson, 1961). Yet by the 1960s, with more information than ever before appearing about the fossil record, fossils seemed merely to be adding uncertainty to phylogenetic reconstruction (Smith, 1984a). Higher taxa were proliferating, being created for poorly understood fossils or for small groups whose unique characteristics were stressed or which did not have all of the characteristics thought essential for placement in established taxa. This practice simply avoided and confused the main issue that had occupied earlier systematists, namely the recognition of relationship. Furthermore, similarity in its broad sense was seen as sufficient evidence for ancestor recognition, particularly when linked with 'evidence' of stratigraphic succession.

Hennig (1966) recognized that fossils could be important for phylogeny if analysed correctly, and suggested that they could potentially be used to determine the polarity of a character or demonstrate instances of convergence. However, he also pointed out that fossils are typically less completely known than recent taxa and thus provide less information for phylogenetic reconstruction.

As Hennig's methodology became widely accepted, the failings of evolutionary systematics and its handling of fossil data started to become glaringly obvious. The primacy of fossils in unravelling phylogeny was challenged from all sides. Patterson (1977) and Nelson (1978b) argued that because fossils could only be interpreted in the light of modern taxa, their role would always be secondary. Engelmann & Wiley (1977), Wiley (1981), Patterson (1981), Schoch (1986), Nelson (1989b) and others demonstrated that ancestors could never be discovered on morphological criteria, and that stratigraphic succession as evidence of ancestry or character polarity was based on an *ad hoc* assumption. In a wide-ranging review, Patterson (1981) claimed that fossils had had virtually no influence in establishing relationships among Recent groups. His explanation for this was that fossils provided many fewer data than living taxa from which to establish phylogenetic relationships. This view was supported by workers such as Smith (1984b), Ax (1987), Meyer & Wilson (1990), and Meyer & Dolven (1992).

This outright attack on the role of palaeontology in phylogenetic reconstruction was not without justification. Methodological approaches for the analysis of fossil data had been sloppy and were certainly in need of a complete overhaul. But largely retrospective reviews such as Patterson's (1981) do not address the question of whether, *given the correct approach*, fossils might not provide important lines of evidence.

In response to such criticism, there has been a much needed reappraisal of fossils in phylogeny which has highlighted the critical role that they can sometimes play. This renewed appreciation of palaeontological data has come about because of advances in three fields:

1 Empirical tests including or deleting fossil taxa from a phylogenetic analysis of a clade with extant members.
2 A refocus of cladistic methods to encompass total evidence.

3 Simulation experiments designed to test under what conditions parsimony will fail to identify the correct tree.

Empirical tests showing the importance of fossils

Although there had been previous studies where the inclusion or exclusion of taxa were shown to influence the outcome of the phylogenetic analysis (e.g. Maddison *et al.*, 1984), it was not until the papers of Doyle & Donoghue (1987) and Gauthier *et al.* (1988) (summarized and extended in Donoghue *et al.*, 1989) that fossils were demonstrated to play a pivotal role in establishing phylogenetic relationships. The approach adopted in both cases was very similar. A major clade with a long fossil record was chosen (seed plants by Doyle & Donoghue; tetrapods by Gauthier *et al.*) and data matrices constructed that included both Recent and fossil taxa. The results from a cladistic analysis of this full data set were then compared with results after removal of some or all of the fossil taxa. When fossils were omitted 'every effort was made to score the characters for the remaining groups as though the excluded groups had never been discovered' (Donoghue *et al.*, 1989, p. 434). In this way it was possible to explore the effect on tree topology of incorporating data from fossils.

In the case of tetrapods it was demonstrated that the inclusion of fossil taxa made an appreciable difference to the topology of the tree. For example, using 109 characters derived only from the Recent taxa Gauthier *et al.* (1988) found that mammals were sister group to crocodiles plus birds, and that Testudines (turtles) were outgroup to those three taxa together (Fig. 3.11). However, when the full data set of Recent and fossil taxa was analysed, mammals were placed as sister group to all other tetrapods, while Lepidosauria (lizards) were placed as sister group to crocodiles and birds (Fig. 3.11).

In the case of seed plants the addition of fossils had little impact on tree topology, but affected the results in other ways. In particular, the addition of fossils tended to make some relationships much more robust and altered the interpretation of a number of character polarities and transformations.

Additional studies have now been carried out that support the idea that, if fossil taxa are ignored, less resolved or positively misleading topologies result. Novacek (1992a,b), Wilson (1992), and Cloutier (submitted) have all shown that ignoring data from fossils can lead to different (and possibly wrong) trees. Novacek (1992a) examined mammal relationships at ordinal level. Although fossils made little appreciable difference to the data matrix (only 7 of the 88 characters required recoding after removal of all fossil taxa), these few differences were enough to cause some topological rearrangement. Parsimony analysis restricted to extant taxa left the position of Hyracoidea unresolved, with two equally parsimonious solutions (Fig. 3.12). When fossil representatives were added to the analysis, hyracoids were ident-

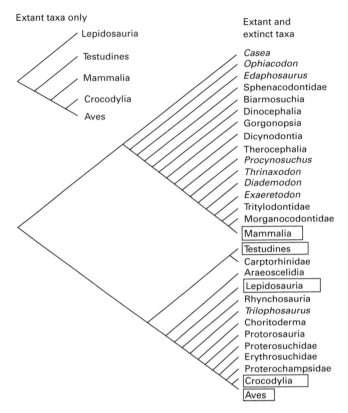

Extant taxa only

Extant and
extinct taxa

Lepidosauria

Testudines

Mammalia

Crocodylia

Aves

Casea
Ophiacodon
Edaphosaurus
Sphenacodontidae
Biarmosuchia
Dinocephalia
Gorgonopsia
Dicynodontia
Therocephalia
Procynosuchus
Thrinaxodon
Diademodon
Exaeretodon
Tritylodontidae
Morganocodontidae
Mammalia
Testudines
Carptorhinidae
Araeoscelidia
Lepidosauria
Rhynchosauria
Trilophosaurus
Choritoderma
Protorosauria
Proterosuchidae
Erythrosuchidae
Proterochampsidae
Crocodylia
Aves

Fig. 3.11 Empirical tests of the importance of fossils: tetrapod relationships
(Gauthier *et al.*, 1988). The cladogram in the top left-hand corner is derived from
characters scored on extant taxa only. The main cladogram incorporates a large
number of extinct taxa in addition to the five extant taxa (boxed) and generates a
different hypothesis of relationships.

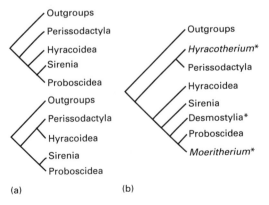

Outgroups
Perissodactyla
Hyracoidea
Sirenia
Proboscidea
Outgroups
Perissodactyla
Hyracoidea
Sirenia
Proboscidea

Outgroups
*Hyracotherium**
Perissodactyla
Hyracoidea
Sirenia
Desmostylia*
Proboscidea
*Moeritherium**

(a)   (b)

Fig. 3.12 Empirical tests of the importance of fossils: hyracoid relationships
(Novacek, 1992a). (a) Shows the relationships when only extant taxa are considered:
there are two equally parsimonious trees. (b) The result when a number of fossil
groups (asterisked*) are included in the analysis: the cladogram is fully resolved.

ified as sister group to Sirenia plus Proboscidea (Fig. 3.12). In this case the addition of fossil taxa has improved resolution.

The relationship of sarcopterygians (lungfishes, coelacanths, and tetrapods) tackled by Cloutier (submitted) is more equivocal because of the general lack of agreement among workers over homologies. Cloutier compiled a data set of 157 characters and 38 taxa and carried out two analyses, one with extant taxa only and a second combining both Recent and fossil taxa. Recent taxa alone identified tetrapods as sister group to lungfishes plus coelacanths, but adding fossil taxa changed the topology, making coelacanths sister group to lungfishes plus tetrapods (Fig. 3.13). Marshall & Schultze (1992) have also pointed out that the interpretation of character transformations in this three-taxon problem is grossly distorted if fossil data are ignored. When fossils are ignored, homoplasy can be misinterpreted as homology.

Total evidence as the most stringent test of homology

There is a strong case for basing cladistic analyses on the total evidence that can be assembled (Kluge, 1989; Eernisse & Kluge, 1993; Kluge & Wolf, 1993). The argument goes as follows:

1 Our estimate of phylogeny is based on identifying a hierarchical pattern among characters believed to be homologous.

2 The best method of testing the homology of characters is by congruence in cladistic analysis (de Pinna, 1991; Patterson, 1982; see above, p. 35): the topology that gives the greatest number of congruent distributions is preferred and all characters that show incongruent distributions are then interpreted as homoplasy. Homology and homoplasy can only be distinguished on the strength of character congruence.

3 The more characters and the more taxa that can be incorporated the more severe a test of homology they provide. Recent taxa only provide us with a small sample of all the character combinations that have existed and the most stringent test of character homology comes from including all taxa in an analysis, both Recent and fossil.

However, the addition of taxa with significant amounts of missing data often leads to a large increase in the number of equally parsimonious solutions (Rowe, 1988; Novacek, 1989, 1992a; Swofford & Olsen, 1990). As fossils are often said to be more incompletely known bacause of missing data (non-fossilization), then possibly their inclusion will merely destabilize results and collapse trees into polychotomies. Kluge (1989) argued that, unless there was a priori evidence that certain kinds of characters would confound the analysis, all should be included. Should fossil taxa with significant amounts of missing data therefore be excluded from analysis?

There are two aspects of missing data that are relevant to this question: (i) completeness and informativeness are not necessarily the same thing (Donoghue et al., 1989; Cloutier, submitted); and (ii) the problem of missing data is not confined to fossils but applies equally to Recent taxa (Gauthier et al., 1988; Donoghue et al., 1989).

Completeness and informativeness of fossils. In assessing the contribution fossil taxa can make to a phylogenetic analysis, it is not the number of characters that can be coded that should be considered, but the relative importance of those characters to the outcome of the phylogenetic analysis. Cloutier (submitted) tested the role individual fossils had to play in establishing sarcopterygian relationships empirically. Firstly he constructed a cladogram of sarcopterygian relationships using a large suite of morphological characters that incorporated both Recent and fossil taxa that were relevant (Fig. 3.13). Then he deleted each fossil in turn from the analysis and noted what effect this had on the cladogram topology. Cloutier found that deleting some fossil taxa had no effect on the results of the parsimony analysis, whereas the deletion of others eliminated character conflict and thus reduced the number of equally parsimonious solutions found. Fossils in this latter group were having a detrimental effect on the parsimony analysis, presumably bacause of the problems associated with missing data. Deletion of taxa belonging to a third group affected only relationships in the immediate vicinity of the tree (i.e. only influenced the

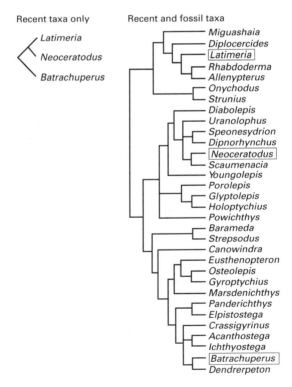

Fig. 3.13 Empirical tests of the importance of fossils: sarcopterygian relationships (Cloutier, submitted). The cladogram on the left is derived from a character matrix of extant taxa. The main cladogram shows the relationships after including 29 fossil taxa. Adding fossil taxa causes lungfishes (*Neoceratodus*) to group with tetrapods (*Batrachuperus*) rather than coelacanths (*Latimeria*).

position of other fossil taxa). Finally, the deletion of a few key fossil taxa influenced the overall topology of the tree in a significant way, altering the relationships of extant taxa. One of the more important fossil taxa in this last category was *Diabolepis*, which could be scored for only 74 of 157 characters (47%), yet these characters prove crucial for settling the position of dipnoans within rhipidistians.

It is therefore clear that certain taxa can still play an influential role in phylogenetic reconstruction even when they have a large proportion of missing data. In such taxa those characters that can be scored are apparently highly informative in phylogenetic terms. It is not justifiable to remove fossil taxa simply because they cannot be scored for a sizeable portion of their characters, contrary to the recommendation of Rowe (1988).

Fossils and missing data. Fossils are less completely known than modern taxa, due to information that has simply not been preserved (i.e. most soft-tissue characteristics); but missing data can also be a significant problem for skeletal aspects of Recent taxa, as highlighted by Gauthier *et al.* (1988). As sister groups diverge, some characters can become transformed beyond certain recognition while others can be lost altogether. Consequently there may be a number of characters that must be scored as missing when dealing with extant taxa. For example, the Aristole's lantern is a complex structure in echinoids that is extremely important in the classification of the group, but in two major derived clades – the heart urchins (Atelostomata) and Cassiduloida – this structure has been lost and all characters associated with it must be scored as missing.

The extent of this problem varies from group to group. Novacek found that only 1% of his characters were uncodable for extant mammal orders, whereas Gauthier *et al.* (1988) reported 15% of their amniote characters uncodable for extant mammals and 19% for extant turtles. Cloutier (submitted) had 41% of characters unscored in modern tetrapods for his sarcopterygian data set. Interestingly, Cloutier was able to show that the number of characters that cannot be scored because of evolutionary transformation increases with age whereas plesiomorphic character states decrease (Fig. 3.14).

If total evidence provides the best way to secure sound phylogenies, there is no justification for excluding fossil taxa. Although missing data in fossils can lead in some circumstances to an increase in the number of equally parsimonious trees, there is no a priori method for deciding which taxa can be deleted without loss of information, because completeness and informativeness are not the same thing. Furthermore, incompleteness through missing data is not confined to fossil taxa: when highly derived modern taxa are included along with plesiomorphic sister groups, they may have a large number of characters that cannot be scored because they are transformed beyond recognition

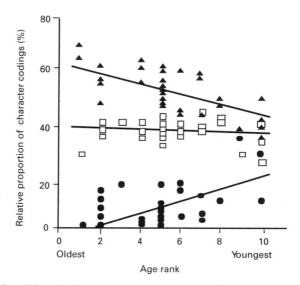

Fig. 3.14 Plot of the relative numbers of plesiomorphic (▲), apomorphic (□) and illogical (●) characters plotted against age rank for sarcopterygian fishes (Cloutier, submitted). The lowest age rank corresponds to the oldest taxa. Regression lines are calculated for each. Note that the number of apomorphic characters (as a proportion of all characters) remains relatively constant, whereas plesiomorphic character states are most common in the oldest taxa and characters that cannot logically be coded for are greatest in the most recent taxa.

Simulation studies

The third line of evidence that has a bearing on the pivotal position of fossils in phylogenetic reconstruction comes from the field of molecular biology. With the revolution in molecular biology and the ability to generate large amounts of nucleotide sequence data rapidly and reliably, there has been a huge surge of interest in methods of constructing phylogenies from molecular data (Hillis & Moritz, 1989; Miyamoto & Cracraft, 1991). The simplicity of the data base, composed of only four nucleotide bases, has advantages and disadvantages. Its main disadvantage is that homoplasy, through overprinting (mutation from one base to another occurring twice at the same position) can become a major problem over time and may swamp any signal of synapomorphies. However, the simplicity of the system makes it relatively straightforward to model, and a large number of workers have used simulation experiments to gauge the success of computer algorithms for tree reconstruction (e.g. Lanyon, 1988; Nei, 1991; Debry, 1992). These studies begin with a known sequence and allow it to 'evolve' (transform) according to set models, so that one ends up with a number of descendant sequences of known relationship. Finally these end sequences are treated as if they were observed sequences from terminal taxa, and the various methods of phylogenetic reconstruction (e.g. parsimony, maximum likelihood, distance matrix methods, etc.)

are applied to see how well they can recapture the known phylogeny.

Using simulation experiments, Felsenstein (1978) first demonstrated that parsimony analysis could fail to find the correct tree if the topology had both long and short terminal branches, since long branches tend to pair together simply through chance match. Subsequent work (e.g. Felsenstein, 1988; Lanyon, 1988; Debry, 1992) has extended and generalized this observation to tree topological considerations. Basically, low stemminess and imbalance of tree branching both increase the likelihood that parsimony analysis will fail to find the correct tree. Stemminess (Fig. 3.15) refers to the relative lengths of internal and terminal branches. If a tree has short internal branches and long terminal branches (for example, if it comprised only extant members of a clade that radiated rapidly in the distant geological past), then homoplasy is likely to swamp any signal of relationship and spurious trees will result, especially if there is any inequality in rate of evolution. Tree balance refers to the extent to which the tree is dichotomous (highly balanced) or pectinate (highly imbalanced) (Shao & Sokal, 1990). A pectinate pattern confers a strong likelihood that adjacent long branches near the base of the cladogram will pair together simply through chance match.

Simulation studies of phylogenetic reconstruction methods in molecular evolution have been criticized as incorporating too many assumptions of unknown validity (Miyamoto & Cracraft, 1991; Allard & Miyamoto, 1992) despite the relative simplicity of the system. For morphological evolution, the situation is hugely more complex and cannot realistically be modelled. Huelsenbeck (1991b) has tried, but

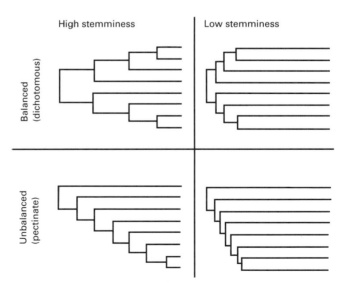

Fig. 3.15 Schematic cladograms to illustrate the difference between balanced and unbalanced topologies (rows) and between cladograms with high stemminess and low stemminess (columns).

his model is very simplified and is therefore of questionable validity. Nevertheless, Huelsenbeck's study suggests that homoplasy of morphological characters in long branches of a tree (through convergence or parallel evolution) could be a significant problem for parsimony analysis. The structure of the tree may thus influence the ability of parsimony to reconstruct the correct topology.

Available evidence therefore suggests that long branches in a cladogram can generate spurious topologies through noise (homoplasy) masking original signal (homologous similarity). It would therefore be prudent to break up such branches by the addition of taxa, so as to create a more balanced topology with more equal branch lengths. For molecular phylogenies this may be impossible – there is only one living representative of coelacanth, for example, and the long branch extending back to its divergence from extant lungfish and tetrapods cannot be subdivided. Similarly, mammals all belong to a crown group that has diverged over the past 140 Ma (Benton, 1990) so it is impossible to find any living tetrapod that can act to break up the long stem branch of this clade. For morphological phylogenies, however, we can turn to the fossil record for taxa that can be added to the data matrix to divide long branches and help distinguish synapomorphy from homoplasy (e.g. Marshall & Schultze, 1992). The addition of fossils that help to break up long branches could therefore be crucial to the success of a parsimony analysis.

Why fossils are important

There is little doubt that, under certain circumstances, fossils can have a crucial role to play in phylogenetic reconstruction when considering characters that are equally recoverable from Recent and fossil taxa. This is probably because the addition of fossils provides a denser sampling of skeletal character assemblages that have existed within a clade than could be gleaned from considering the living biota alone. Denser sampling improves our chances of recognizing homoplastic similarity in terminal branches for what it is. The denser our sampling, the closer we get to total evidence (though of course we shall always be estimating from incomplete data) and the better our chances of achieving the correct topology. The addition of fossils will improve our knowledge of homology in skeletal characters.

Nevertheless, it is still true that the addition of other kinds of characters (biochemical, histological, genetic) that can only be gleaned from living taxa will ultimately prove more crucial than fossil data, since it is the addition of more putative homologies which provides the key to developing more robust phylogenies.

Within the context of skeletal-based phylogenies, fossils will have a more significant effect if the tree includes long terminal branches and short internal branches, or if there are relatively few extant taxa to sample. Where the crown group has a long stem group, fossils will also play an important role in ordering morphological changes that have occurred in the stem group. Furthermore, although the addition of any taxon has the potential to change the topology of the tree because of its character combination, only fossils have the potential to break up long branches.

Adding fossil (or Recent) taxa can alter cladogram topology in one of three ways (Fig. 3.16):

1 It can reveal that a character treated as a putative synapomorphy in two of three groups is actually present in the fossil antecedents of the third group. This identifies the derived character state as being

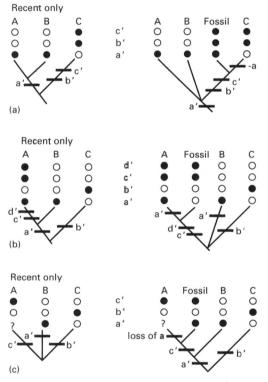

Fig. 3.16 Schematic cladograms showing the three possible ways in which adding taxa (in this case, fossil) can alter the interpretation of character state changes. Recent taxa A, B, and C are shown along with a character matrix. (a) The addition of a fossil taxon shows that character a′ is more general in distribution than had been suspected from Recent taxa alone. The character becomes a synapomorphy at a lower level in the cladogram and is reversed higher up. (b) The addition of a fossil taxon shows that character a′ has arisen more than once and that a homoplastic similarity has been mistaken for homology in Recent taxa. The character becomes two synapomorphies in more terminal portions of the cladogram. (c) The addition of a fossil taxon shows mixed character states. In this case taxon A cannot be scored for character a′ because the structure to which that character relates is missing in that taxon (for example it has been lost or transformed beyond recognition). The addition of a fossil taxon with synapomorphies linking it to A (character c′) but which also show the character state a′ that is primitive for clade (A+B), allows resolution of relationships.

more general than previously suspected and changes the polarity of that character in the context of the three-taxon problem. Garstang (1928), for example, resolved the relationship between chordates, hemichordates, and echinoderms on the basis of comparative anatomy of modern forms. He observed that both hemichordates and chordates have gill slits and that these are not found in extant echinoderms or in any other group. From this he argued that hemichordates and chordates were the more closely related. However, Jefferies (1986) has shown that there are some fossils (cornutes) which have unambiguous ex- halant gill openings. These cornutes also have a unique characteristic of echinoderms, namely the stereom structure of their skeleton. This implies that gill slits were common to primitive echinoderms and thus removes the evidence for hemichordate–chordate relationships. The presence of gill slits becomes a character supporting a deeper node in the cladogram, and it is the loss of gill slits in the lineage with a stereom skeleton (echinoderms) that is revealed as the derived state.

The same effect occurs if the presumed derived state shared by two of the three taxa is shown to be present in the stem group of all three. In both cases the presumed synapomorphy is shifted root-ward to a deeper node.

2 The addition of fossil taxa can lead to an assumed homology between two clades being identified as having arisen through convergence. This occurs when a fossil taxon of one or other lineage (i.e. with at least one unambiguous autapomorphy of these clades) is discovered which has the plesiomorphic condition for this trait (Fig. 3.16b). The distal phalanges (fingers) in perissodactyls (horses) and Hyracoidea show a broadly similar derived condition which would be scored as a putative homology on the basis of living groups alone (Novacek, 1992a). How- ever, uncontested fossil perissodactyls such as *Hyracotherium* are found to lack the derived state of the distal phalanges, which must be inferred to have evolved through convergence in the two groups.

The hypapophyses on the anterior trunk and posterior cervical vertebrae of Recent crocodiles and birds provides another example (Gauthier *et al.*, 1988). Considering living amniotes alone, this de- rived similarity might be treated as homologous and a putative syn- apomorphy for birds and crocodiles. However, when fossils are considered it is evident that the stem group taxa of both crocodiles and birds lacked hypapophyses, which therefore must have evolved inde- pendently in the two lineages. Marshall & Schultze (1992) provide further examples of convergence between lepidosirenid lungfishes and tetrapods, as compared to the lungfish *Neoceratodus*. In all these cases, when a putative homology is inferred to be a homoplasy the character becomes two or more (convergent) synapomorphies posi- tioned further towards the termini in the cladogram. The addition of fossil data again acts to collapse relationships.

3 The addition of fossil taxa from the stem lineage of a highly derived extant group (i.e. a fossil with some, but not all, of the synapomorphies of the extant group) can make a positive change if some characters in extant members are transformed beyond certain recognition or lost altogether. In a three-taxon problem, this is equivalent to having a character scored as separate states in two of the taxa and as unknown in the third (Fig. 3.16c). A fossil that is a member of the stem group may reveal homologous characters in a less transformed state, allowing them to be scored and thus providing additional synapomorphies. This appears to be the main reason for the marked change in topology among tetrapod clades when fossil synapsids (stem group mammals) are added (Gauthier *et al.*, 1988). For example, post-dentary bones are so transformed in extant mammals (as ear ossicles) that their structure is impossible to compare in detail with that of other tetrapods. In synapsids these bones are less transformed and consequently more closely comparable with those in members of other stem groups. Adding synapsids to an analysis allows post-dentary characters to be scored as synapomorphies in the line leading to extant mammals.

As Novacek (1992a,b) pointed out, fossils are likely to have the most effect if there is already a character conflict concerning a taxon's position. Adding fossils to an analysis can only result in one or more synapomorphies switching their focal level in the cladogram. Where branches are poorly supported this can then be sufficient to swing the balance of evidence towards an alternative topology.

It seems most unlikely that fossils will ever be able to generate sufficient evidence on their own to identify relationships that have no support whatsoever from comparative anatomy of extant forms alone. Therefore the recognition of suites of additional synapomorphies through the analysis of non-skeletal characters among extant taxa is always likely to be more important to phylogenetic analysis than the addition of fossil taxa, which in reality only fine-tunes existing synapomorphy schemes.

Summary

Cladistic methodology provides the best approach for establishing phylogenetic relationships among taxa. The concept of homology lies at the core of comparative biology and the cladistic method. Characters are statements – about homology based on observed similarity – which can be tested, using the criterion of congruence, by other statements of homology. The greater number of putative character homologies that can be identified, the greater the test they provide of any single homology statement.

The construction of a cladogram proceeds from the compilation of a character–data matrix. Choice and definition of characters is crucial to any analysis and requires careful consideration. There seems little justification for differentially weighting characters prior to a cladistic analysis, but a posteriori weighting based on heuristically determined character consistency indices offers an appropriate means of choosing from among multiple equally parsimonious solutions.

Character states are best left unordered and the cladogram rooted using outgroup taxa. The outgroup method requires that we have a pre-existing hypothesis of ingroup–outgroup relationships to enable selection of the most appropriate outgroup taxa. This is not a major drawback, because all methods build on existing scientific knowledge and it is not unreasonable to assume that we have got somewhere in the past few hundred years of systematic research (Eldredge, 1979).

Outgroup rooting works best when two or more outgroups are used and when the outgroups are the closest known relatives to the ingroup. The latter point is important in that it explains the major role of fossil taxa. The inclusion of (i) ingroup fossils that lie close to the basal node of the ingroup and (ii) outgroup fossils that lie close to the node separating outgroup from ingroup, enhances the chances of obtaining the correct topology. Outgroup rooting also has the advantage that character polarities do not have to be set prior to a numerical cladistic analysis.

Fossil data can play an important role in phylogenetic reconstruction, as has been demonstrated from empirical and simulation tests. Their importance arises not because of any special properties that fossils might or might not have, but because total evidence offers the best test of homology. The additional information on character combinations that can be gained from considering total (i.e. fossil and Recent) morphological evidence rather than simply that derived from Recent taxa can prove crucial, even though fossils are usually much more anatomically incomplete. Ultimately, however, phylogenetic relationships are tested and made more robust through increasing the number of phylogenetically informative characters that can be scored. Fossils potentially offer only very limited access to new characters, whereas a wealth of anatomical, biochemical, and genetic characters is potentially available from extant organisms. For this reason, fossils will always play a subdominant role in establishing phylogenetic relationships.

4 Higher taxa

The foregoing chapters have discussed the meaning of 'species' in the fossil record, and how the phylogenetic relationships of such phena can be established through cladistic analysis. It is now appropriate to consider how these phena are grouped together to form higher taxa, and what effect this grouping process may have on the perception of evolutionary patterns.

Why higher taxa are needed in evolutionary studies

Although raw phenon data are occasionally used to examine broad evolutionary patterns (e.g. Hart, 1990), phena are usually considered to have too patchy and serendipitous a fossil record to provide a secure base for documenting evolutionary patterns. Of course, some taxa do appear to have excellent phenon-level fossil records, such as the open ocean planktonic foraminifer *Globorotalia* (Malmgren *et al.*, 1983), and Paul (1982, 1985) has made the case that the fossil record is not as bad as is often portrayed.

However, it is important not to confuse local with global completeness. Local completeness, as measured by observed phenon distribution within stratigraphic sections, is often very good, at least for the more common phena. But the adequacy of available stratigraphic sections to document the total range of a phenon is more difficult to assess. The range of a phenon may be very well documented in a series of stratigraphic sections, such that we can have great confidence in its local range. If, however, its first appearance represents an immigration from a region where no record of the appropriate facies is preserved, then the observed range may greatly underestimate its total range. In this case the fossil record accurately portrays the distribution of the phenon in the observed sections, but the available sample of sections does not give an adequate representation of the phenon's total range.

There is no doubt that the great majority of phena are rare in the fossil record (Koch, 1991) and thus pose significant problems for sampling. A survey of post-Palaeozoic fish and echinoid phena found that approximately two-thirds (68%) were recorded from only a single locality or local region (Smith & Patterson, 1988). Thus, patterns of phenon diversity are likely to be strongly biased by the distribution of appropriate facies that favour preservation of high diversity faunas. In order to overcome these problems of sampling, many workers (e.g. Sepkoski, 1978, 1984; Raup & Boyajian, 1988; Valentine, 1990; Skelton *et al.,* 1990) have preferred to group phena into higher taxa in order to study evolutionary patterns. The chances of finding a single phenon in beds of the appropriate facies depend on factors such as its preservation potential and its relative abundance, and may be quite low. However, grouping phena together into higher taxa such as genera and families obviously improves the chances of observing a member of that taxon in any one time interval. The more phena included in the taxon, the better the chances that it will be sampled (if present) at a particular level.

Higher taxa are therefore used as proxy for phenon-level data and the evolutionary patterns they display are extrapolated back to the phenon level by inference. Families and genera are used most often as proxies since patterns of extinction and diversification at these taxonomic levels have been observed to track phenon-level patterns (Sepkoski *et al.,* 1981).

Grouping phena thus has advantages in overcoming some of the problems associated with patchiness of the fossil record. However, the way phena are grouped together is important, since it determines whether taxa have objective or conventional boundaries.

The construction of higher taxa

There are only three ways in which phena can be grouped into higher taxa on the basis of morphological similarity (Fig. 4.1).

Monophyletic taxa

Phena can be grouped solely on the basis of the derived homologous characters (synapomorphies) that they share. The presence of a derived morphological character or characters is used as the diagnostic similarity by which members are recognized. In evolutionary terms these derived homologous characters represent morphological innovations inherited by all the descendants of the phenon which first displayed this character. For example, the presence of feathers is a clear and unambiguous derived morphological character that unites all living birds; having feathers is diagnostic of group membership. These characters are *synapomorphies* and the groups thus defined are termed *monophyletic.*

Monophyletic taxa are often defined on their presumed history, i.e. a group which contains an ancestor and all its descendants (Hennig, 1966; Schoch, 1986). However, this is an inference based on the observed distribution of characters among taxa; an operational definition is both simpler and more practical, and therefore preferable. In operational terms a monophyletic group is a group united by one or more synapomorphies on a cladogram (Farris, 1974, 1991). Note that the diagnostic synapomorphies do not necessarily have to be present in

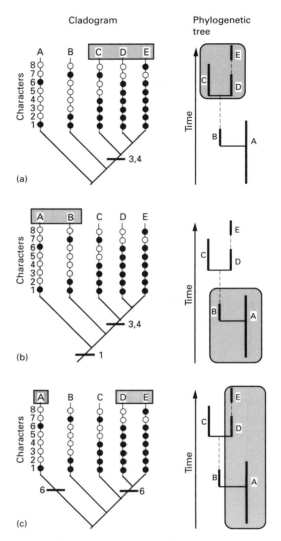

Fig. 4.1 Different ways of constructing taxa, each illustrated by cladogram (left) and phylogenetic tree (right). Taxa being grouped are shown in stippled boxes. The characters used to support the various groups are indicated by bars on the branches of the cladograms. (a) Monophyletic group: based on presence of characters. (b) Paraphyletic group: based on presence and absence of characters. (c) Polyphyletic group: based on characters that have been derived twice independently.

all of the members of the group, but may be reversed in more terminal portions of the cladogram. Secondary loss of synapomorphies within a derived subgroup is not a problem, provided there are sufficient other characters to show that the subgroup nests within the clade defined by those synapomorphies. For example, echinoids differ from other classes of echinoderm in their possession of a lantern apparatus and this forms an excellent synapomorphy for the group. Yet one group of echinoids – heart urchins (spatangoids and holasteroids) – lacks this structure. From the many other characters heart urchins share with lantern-bearing echinoids, it is obvious that the lantern in heart urchins has been lost secondarily.

Monophyletic groups are defined by a single evolutionary event: the acquisition of an evolutionary novelty which marks their inception.

Paraphyletic taxa

Phena can be grouped on the basis of a combination of both the presence and absence of derived homologous characters that they share. These groups are identified by at least two characters: the base of the group is defined by the acquisition of a derived character, while its upper limit is defined by the acquisition of a second character (Fig. 4.1b). The absence of the second derived character is thus treated as equally important as the presence of the first derived character in defining group membership. Such groups are termed *paraphyletic*.

A paraphyletic group is what remains after one or more monophyletic subsets have been abstracted from a more inclusive monophyletic group. They are grades, not clades. Inarticulate brachiopods are a paraphyletic group, composed of phena within the phylum Brachiopoda that have a bilaterally symmetrical, bivalved shell and lophophore (derived characters) but which lack the synapomorphy by which articulate brachiopods are identified, namely an articulatory peg-and-socket system along the hinge-line of the shell (Hennig, 1966; Popov *et al.*, 1993). The inarticulate brachiopods are the remainder of the phylum Brachiopoda after the Articulata has been extracted.

In evolutionary terms a paraphyletic group may be thought of as a group that includes a common ancestor and some, but not all, of its descendants (Oosterbrook, 1987). Operationally, however, it is identified, and should be defined, on the basis of character attributes (Farris, 1991). *A paraphyletic group is one supported by neither a synapomorphy nor a character which shows multiple derivations in a cladogram.*

Paraphyletic groups are a particular problem because their boundaries are set by convention. Consider the example of five phena, A–E, whose character distributions indicate the relationship shown (Fig. 4.2). If each node is supported by a derived character state then there are four monophyletic groups that can be recognized: [A+B+C+D+E], [B+C+D+E], [C+D+E], and [D+E]. There are six additional paraphyletic groups: [A+B+C+D], [A+B+C], [B+C+D], [A+B], [B+C], and [C+D]. The monophyletic groups form a single coherent nested set of taxa, each compatible and uncontradicted by any of the other monophyletic groups. They are thus defined in a totally consistent manner based solely on set criteria. The paraphyletic groups, on the other hand, form an overlapping patchwork of sets. Phenon B, for example, could belong to either [A+B+C] or [B+C+D], which are mutually incompatible. Thus a choice has to be made as to which paraphyletic groups are going to be recognized from among the alternatives. In the past this has been done (usually imprecisely) on the basis of morpholo-

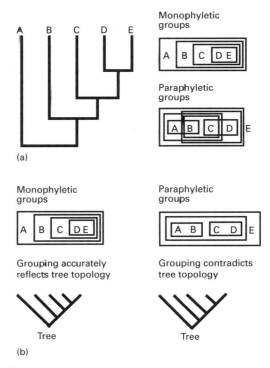

Fig. 4.2 For the five-taxon cladogram illustrated there is only a single, entirely self-consistent nested set of monophyletic groups that can be recognized. In contrast, six mutually contradictory paraphyletic groups can be defined. Monophyletic groups accurately capture tree topology; paraphyletic groupings do not.

gical disparity: if [A+B+C] is perceived to be separated from [D+E] by a considerable number of morphological attributes (i.e. accumulated synapomorphies) then group [A+B+C] is preferred over group [B+C+D].

A phenon can be grouped into a large number of incompatible higher paraphyletic taxa; therefore which scheme is ultimately preferred is clearly dependent upon taxonomist's dictum, and thus imposed by the observer. Monophyletic groups, on the other hand, constitute only a single, completely consistent, nested hierarchy; the recognition of monophyletic taxa is therefore not subject to the judgement of taxonomists. Ball (1975) and Patterson (1982) have used this feature to argue that monophyletic taxa are real and may be discovered through the recognition of homology. By contrast, paraphyletic taxa are groups that have been selected from a set of mutually exclusive possibilities by taxonomists, and are therefore imposed on an *ad hoc* basis.

Polyphyletic taxa

Phena can be grouped on the basis of non-homologous characters. These groups are based on similarities that have arisen through convergence (Fig. 4.1c). This occurs when group membership is defined using a morphological character originally thought to be homologous, but which is later shown to have originated two or more times independently. Such groups are termed *polyphyletic*. In evolutionary terms they represent groups based on mistaking convergence for homology. As with paraphyletic groups, they are created through taxonomic convention and are artificial constructions. For example, they are commonly used in the construction of taxonomic keys and arise when only single characters are relied upon.

An example of a polyphyletic group is the Graptoloidea. Until Fortey & Cooper (1986) examined the question of graptolite phylogeny from a cladistic stance, the Graptoloidea included all taxa without bithecae. This definition specifically excluded members of the Anisograptidae, which had formerly been classified with the Dendroidea. However, when Fortey & Cooper took careful stock of the morphology of Anisograptidae, they found evidence that loss of bithecae may have occurred several times, namely that different primitive members of the Graptoloidea had different anisograptids as sister taxa (Fig. 4.3).

Monotypic taxa

Monotypic taxa contain only one phenon. These higher taxa are not used to group phena, but to make a statement about perceived morphological disparity of the phenon included, or about its sister group relationships. Such taxa are dealt with more fully below, where the question of rank is considered.

What defines a taxon?

Taxa have traditionally been recognized and defined on the basis of the characters of their members. Virtually all formal palaeontological

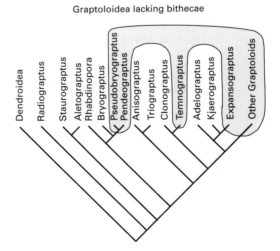

Fig. 4.3 Graptoloidea, when defined on the absence of bithecae, are a polyphyletic group (Fortey & Cooper, 1986).

definitions are based on this principle. However, simply enumerating the synapomorphies that characterize a node in a cladogram is not enough to define a taxon, since reversal higher in the cladogram can result in a subset of phena that lack those character states deemed diagnostic. Thus, characters alone are neither necessary nor sufficient to define a taxon (Sober, 1988). Clearly a taxon is a statement about membership, not character attributes – the characters are diagnostic of a taxon, not defining (Ghiselin, 1984a; Sober, 1988; de Queiroz & Gauthier, 1990).

To define the composition of a taxon it is therefore necessary to state the phena that are included (Ghiselin, 1984a). This can be done in one of two ways:

1 An operational definition uses cladistic structure, derived from formal character analysis, to identify the members of monophyletic taxa. A taxon can then be defined simply by listing the subsets of taxa that it contains at the next level in the hierarchy.

2 A process-related definition, such as that proposed by de Queiroz & Gauthier (1990), defines taxa as 'the most recent common ancestor of [named subgroups] and all its descendants'.

Since common ancestry can only be inferred from cladistic structure and has no independent empirical basis, the operational definition is clearly preferable.

Status of our current taxonomic database

Although it is now generally recognized that only monophyletic taxa should be erected to group phena, this is a relatively recent realization stemming from the work of Hennig (1966). The vast taxonomic database that we have inherited from the systematic endeavours of the past 200 years comprises a chaotic mixture of monophyletic, paraphyletic, polyphyletic, and monotypic taxa, and the work of transforming this into a consistent set of monophyletic taxa has hardly begun. A few groups (e.g. higher vertebrates, arthropods, and echinoderms) are now better studied, and the eradication of paraphyletic and polyphyletic taxa is further advanced. However, many groups remain largely unrevised and the major palaeontological works of taxonomic reference, such as *The Treatise on Invertebrate Paleontology* (Moore *et al.*, 1953–86) and Harland *et al.* (1967), all predate the cladistic revolution.

In a review of the taxonomic status of post-Palaeozoic echinoderm and fish families, Smith & Patterson (1988) found that only a third of families (33%) listed in the most up-to-date (at that time) summary of our taxonomic database (Sepkoski, 1982, plus updates) represented demonstrably monophyletic groups. Of the remainder, 14% were paraphyletic and 21% polyphyletic. A further 21% were monotypic and the rest (11%) were non-monophyletic taxa of uncertain status (Table 4.1). The frequency of monophyly in generic data was about the same (32%), but paraphyletic genera were more common (23%) and

Table 4.1 Estimated composition of traditional taxonomic databases in terms of their monophyletic and non-monophyletic taxa. Three databases were examined for fish and echinoderm entries: Van Valen's (1973a) source data for family level taxa; Sepkoski's published compendium of marine families (1982, with updates to 1986); and Sepkoski's unpublished generic level compendium (every fifth entry checked) (Patterson & Smith, 1987; Smith & Patterson, 1988). Indeterminate taxa (indicated by asterisk*) correspond to taxa not supported by derived character states but whose status as polyphyletic or paraphyletic is unclear. The family database of Sepkoski (1986) also includes 4% non-marine entries

| | Families | | Genera |
	Van Valen (1973a)	Sepkoski (1986)	Sepkoski (1986)
Total number of taxa examined	63	144	305
Monophyletic taxa			
Range correct (%)	19.0	25	32
Range since changed (%)	14.3	15	–
Total (%)	33.3	40	32
Non-monophyletic taxa			
Paraphyletic (%)	20.6	20	23
Polyphyletic (%)	14.3	8	7
Monotypic (%)	20.6	17	12.5
Indeterminate* (%)	11.1	14	25
Total (%)	76.6	59	67.5

polyphyletic genera less common (7%). Monotypic genera accounted for 12.5% of records and non-monophyletic genera for 7%. The remainder (18%) could not be categorized because of a lack of recent revisionary work. If this is generally true for other groups, then only about a third of all higher taxa currently recognized are monophyletic.

Higher taxa and evolutionary patterns

When higher taxa are used to investigate evolutionary patterns, one of two assumptions has to be made: (i) higher taxa represent meaningful biological groupings defined either in terms of their genealogical or morphological cohesiveness; or (ii) higher taxa are used as a means of sampling phenon-level data. Each assumption will be considered in turn.

Higher taxa as meaningful biological entities

If the appearance, disappearance, and duration of higher taxa are to be used to deduce evolutionary patterns, these higher taxa must have some reality in the biological world. The biological basis of monophyletic groups is in no doubt. As argued above, monophyletic groups are based on homology and constitute a single, fully consistent hierarchical framework of taxa. Their biological reality comes from the fact that they represent complete branches of the evolutionary tree. Analysis of their distribution through geological time thus gives information on the geometry of the evolutionary tree.

Polyphyletic taxa are equally uncontentious, since they are created by error when a character that has arisen two or more times independently is mistaken for a synapomorphy. These clearly have no biological reality and can have no place in the study of evolutionary patterns.

The case for paraphyletic taxa, however, is more contentious. Paraphyletic taxa are selected by taxonomists from a number of possible alternative groupings, therefore the nested pattern of taxa generated by grouping paraphyletic taxa is incongruent with the inferred genealogy of the group. The nested pattern of paraphyletic taxa deviates to a greater or lesser extent from the genealogical hierarchy (Fig. 4.2). Paraphyletic groups do not therefore represent meaningful biological groups through accurately depicting genealogical history and tree geometry.

However, many workers (e.g. Van Valen, 1978, 1984, 1985; Bottjer & Jablonski, 1988; Jablonski & Bottjer, 1990; Raup & Boyajian, 1988; Valentine, 1990; Skelton, 1993) have argued that paraphyletic groups can be considered as meaningful biological units because they represent groups of phena occupying a similar 'adaptive niche'; paraphyletic taxa are seen as 'parts of phylogenies delimited adaptively, as inferred from changes in morphology. They are thus natural taxa' (Van Valen, 1985, p. 99). Bottjer & Jablonski (1988, p. 541) argued that paraphyletic groups are useful because they represent 'functionally and ecologically distinct forms' and that such groups can provide useful insights into the 'histories of interesting adaptations'. Raup & Boyajian (1988, p. 113) further argued that 'the paraphyletic genus serves effectively to define a group of phena occupying a small portion of morphospace'.

The problem with these arguments is that, effectively, the systematist is left to decide what is 'adaptively unified'. There seem to be no criteria by which such groups can be discovered objectively. As Patterson (1981, p. 207) pointed out, no one can argue that Invertebrata (a paraphyletic group) are adaptively unified, and the same is true for the great majority of taxa. Indeed, it is questionable whether any group above species level can be considered to have ecological attributes (Wiley, 1981; Eldredge & Salthe, 1984; Eldredge, 1985b). Therefore there is clearly no universal correlation between paraphyletic groups and 'adaptive unification', and any such link must be argued individually. This suggests that paraphyletic groups have 'adaptive unity' only in the mind of the taxonomist.

As for the argument that paraphyletic taxa represent small regions in morphospace and can therefore be considered as valid units of analysis (Sepkoski, 1987; Raup & Boyajian, 1988; Valentine, 1990), this misses the point that the boundaries of such regions are, with very few exceptions, set by default and thus arbitrary. A phenon belongs to just one nested series of monophyletic groups but a plethora of mutually incompatible series of paraphyletic taxa. Its allocation to a single paraphyletic taxon is therefore a taxonomist's decision. There are quantitative methods of morphometric analysis that could be applied to cluster phena objectively in morphospace (e.g. Cheetham, 1986, 1987; Jackson & Cheetham, 1990; Tabachnick & Bookstein, 1990; Budd & Coates, 1992; Foote, 1991a,b,c; Gould, 1991) but such techniques have rarely been applied. Instead, paraphyletic groups have almost always been created by default (as a group containing phena left over after monophyletic subsets have been abstracted from a clade) rather than design (multivariate morphometric analysis and numerical taxonomy). As a consequence, there is no reason to believe that existing paraphyletic taxa should accurately depict phena distributions in 'morphospace'.

Raup & Boyajian (1988) believed that paraphyly is a more serious problem at higher taxonomic levels, and argued that by using genera rather than families or orders, many of the problems associated with paraphyly could be avoided in evolutionary studies. However, Smith & Patterson (1988) found comparable levels of monophyly and paraphyly in genera and families. Furthermore (as argued in Chapter 2) paraphyly remains a major problem even at the level of the phenon.

A case study: paraphyly and rudists

The view that paraphyletic groups represent ecological units is so pervasive that it is worth examining one such case in detail. Skelton *et al.* (1990) argue for paraphyletic taxa as 'a clear ecological grouping of related species', citing an example drawn from rudist bivalves (Hippuracea) in support of their claim. Rudists are a monophyletic group of often bizarrely-shaped, sessile, epifaunal bivalves that flourished in shallow tropical seas from the late Jurassic to late Cretaceous. Skelton (1985) had divided rudists into two groups: those with an invaginated ligament (a monophyletic clade) and those without. Possession of an external ligament is a character common to all other bivalves, so the latter group is clearly paraphyletic. Skelton *et al.* (1990, p. 94) argued that 'the survival or otherwise of such a paraphyletic taxon . . . registers valuable information about the fortunes of a certain adaptive type vis à vis its modified descendant stocks and so it is not necessarily a mere taxonomic artefact'. Their point seems to be that useful comparisons can be drawn from the histories of the two groups, since rudists with an internal ligament are constrained to grow in a different manner from those with an external ligament.

The initial point – that rudists lacking an invaginated ligament are ecologically defined – can be quickly dispelled. Skelton (1985, p. 160) categorized rudists into three 'broad adaptive zones': encrusters, recumbents, and elevators (Fig. 4.4). Rudists with an external ligament were categorized as 'essentially' encrusters (although some taxa such as *Toucasia* and *Requienia* are best described as facultative elevators). Rudists with an internal ligament or no ligament were categorized as encrusters, facultative elevators, elevators, and recumbents. Thus, even at this level of comparison the two groups clearly overlap in Skelton's scheme.

What then about the question of the external ligament restricting rudists to spiral growth, while those with an invaginated ligament are unconstrained and can uncoil? Does this generate 'ecologically distinct partitioning'? Again, there are no clear boundaries because some forms with an external ligament can achieve a considerable degree of un-coiling and elevation by umbone projection (e.g. *Macrodiceras*), whereas others with an internal ligament are encrusters that grow spirogyrally (e.g. *Gyropleura*).

Skelton (1985) believed that the evolution of an invaginated liga-ment was a significant event, a key innovation in rudist evolution that triggered the group's rapid diversification. Assessment of whether this was indeed the case is hampered by the lack of any cladistic analysis of rudist relationships. Only a few genera have as yet been treated cladistically (Skelton, 1991). However, Skelton (1978, 1985) provided an evolutionary grade tree (Fig. 4.4) and discussed sufficient character evolution for a preliminary cladogram to be constructed.

In summary, the most primitive group of rudists recognized by Skelton is the Diceratidae. Most attach to the substratum by their left valve (LV), but a few attach by their right valve (RV). The Heterodi-ceratinae attach by their LV whereas the Diceratinae attach by their RV. Skelton derives two groups from the Diceratidae, both of which show much more strongly pronounced inequality of valves. One group, the Requieniidae, attach by their LV and have an operculiform RV. The other, the Caprotinidae, attach by their RV and have a reduced LV. The earliest members (*Valletia*) of the Caprotinidae have an external ligament, but all other genera have invaginated ligaments. Furthermore, some of those with invaginated ligaments have a porous, honeycomb-like wall at points of muscle attachment, while others have solid walls. Skelton derives three other families from this group: (i) the Radiolitidae have a *Monopleura*-type hinge articulation and a porous wall structure; (ii) the Caprinidae have a *Pachytraga*-type peg-and-socket hinge articulation and extensive pallial sinuses in the shell wall; and the Hippuritidae have compact shell walls and internal pillars and also represent a monophyletic group. Using these obser-vations it is possible to reconstitute Skelton's evolutionary tree as a cladistic hypothesis (Fig. 4.4).

From this cladistic hypothesis it is now clear that Skelton's division of rudists – those with an external ligament and those with an in-vaginated ligament or no ligament – is only one of several possible divisions. For example, rudists could be divided into those: (i) attached

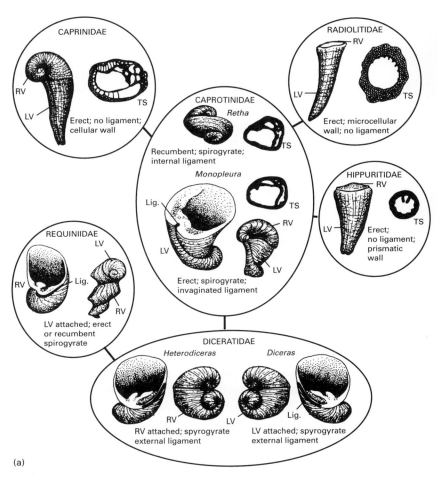

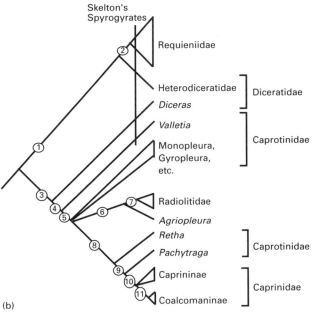

(a)

(b)

by their LV or by their RV; (ii) with or without pallial canals; or (iii) with or without a *Pachytraga*-type moyophore, etc. Each division represents an evolutionary innovation that could have been important in determining the success or otherwise of the sister groups. The important question is, then, How does the derived sister group fare in comparison with its closest relative that lacks the derived trait? The danger with generalizing about an arbitrarily defined paraphyletic group is that the more distantly related members to the derived group will differ in many more traits than the one under investigation, clouding the comparison. By comparing only sister taxa, we can focus on the evolutionary outcome of changing just the smallest possible number of morphological variables at any one time. Sister taxa show the closest morphological similarity, and thus are the most appropriate for comparison.

Furthermore, any study that purports to examine 'key innovations' needs to take a hierarchical approach, examining the significance of trait acquisition at each node. Skelton (1985) provided documentation for the number of rudist phena described from the Kimmeridgian to Aptian, and used this to compare the relative 'success' (in numbers of phena) of his two groups. He observed that rudists with internal ligaments were more diverse than those with external ligaments and concluded that ligament invagination was the key event that liberated rudist evolution. However, this is insufficient to demonstrate that the imbalance of phenon diversity between the two groups is related to that particular node of the cladogram, since arbitrary partitioning of rudists at any of the adjacent nodes would produce the same inequality. Only by testing the relative success of each clade against its sister group is it possible to identify those morphological innovations that might be assigned greater significance. 'Key innovations' must be tied to the correct level in the hierarchy, i.e. the node where a significant change in taxonomic diversity commences (Jensen, 1990; Taylor & Larwood, 1990; Sanderson & Bharathan, 1993). The alternative – to argue on an *ad hoc* basis for a 'macroevolutionary lag' before the effects of any trait became manifest in diversity changes – simply opens the floodgates to rank speculation.

Fig. 4.4 (*Opposite.*) Evolutionary tree and derived cladogram for late Jurassic and early Cretaceous rudist groups. (a) The evolutionary tree is derived from data given in Skelton (1978, 1985, 1991) and shows his division into spirogyrate and uncoiled grades. Representative taxa are illustrated. TS, transverse section through lower valve; LV, left valve; RV, right valve; Lig., ligament groove or pit. (b) A tentative cladogram constructed from Skelton's evolutionary tree. Characters 1–11 are as follows: 1, attached to substratum by RV: single tooth in LV, double tooth in RV; 2, siphonal bands present; strongly inequivalved with cap-like LV; 3, attached to the substratum by LV, RV with single tooth; 4, strongly inequivalved; cardinal platform for anterior adductor muscle attachment; 5, internal ligament; 6, cellular outer shell layer; short massive apophyses; 7, myophore apophyses well developed; 8, pachytragiform myophore arrangement; 9, accessory cavities in inner shell layer; 10, accessory cavities developed into pallial canals in LV; 11, accessory cavities transformed into pallial canals in RV.

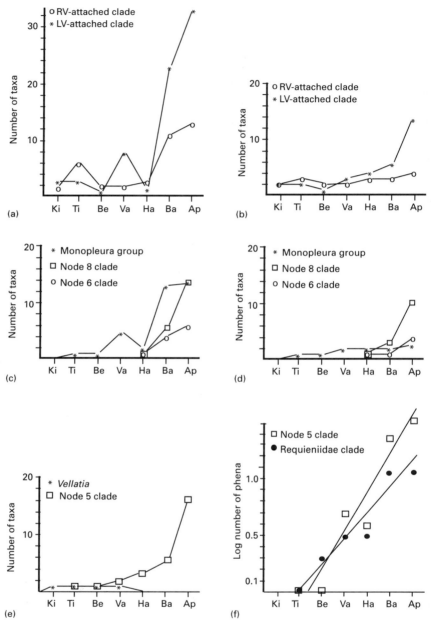

Fig. 4.5 (*Above and opposite.*) Taxonomic diversity for Kimmeridgian to Aptian rudists (Skelton, 1985) placed in a cladistic framework. (a) Number of phena for RV-attached clade versus LV-attached clade. (b) Number of genera for RV-attached clade versus LV-attached clade. (c) Number of phena for the three clades defined at node 5. (d) Number of genera for the three clades defined at node 5. (e) Number of genera in the two clades defined by node 4. (f) Number of phena for the Requieniidae and the clade defined by node 5, plotted on a log scale. (g) Successive clades plotted as a stacked histogram for LV-attached clade. (h) Diversity of RV-attached clade for comparison. Ki, Kimmeridgian; Ti, Tithonian; Be, Berriasian; Va, Valanginian; Ha, Hauterivian; Ba, Barremian; Ap, Aptian.

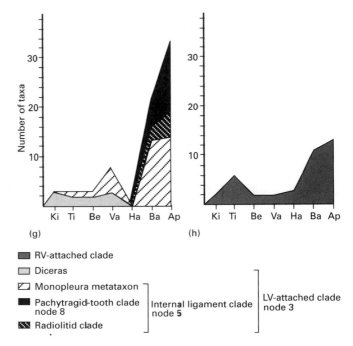

Fig. 4.5 *Continued.*

Figure 4.5 represents Skelton's (1985) data in the form of clade comparisons, as far as is possible. The first comparison that can be made is between rudists attached by their LV with those attached by their RV (Fig. 4.5a&b). This split had occurred by the Kimmeridgian, yet, both at generic level and in terms of total phenon diversity, the two clades have broadly comparable diversity until the Barremian. From the Barremian onwards, however, the LV-attached clade starts to become very much more diverse and by the Aptian the two groups differ significantly. Since no appreciable difference in diversity is apparent prior to the Barremian, the assumption must be (as Skelton concluded) that whether attachment was by the LV or RV had no effect on rudist diversity.

The development of pronounced valve asymmetry separates both LV and RV diceratids from their respective sister groups. In both cases the derived sister group has become more diverse. Thus there may be grounds for suggesting that this is a significant morphological acquisition. By contrast, the acquisition of an invaginated ligament does not create any immediate differential diversity: internal ligaments first appear in the Tithonian, but diversity in the LV clade does not change until the Berriasian.

When the diversity of individual nested clades is compared (Fig. 4.5g&h) it becomes clear that the appearance and rapid diversification of two clades – the Radiolitidae and the pachytragid clade – coincides with the dramatic increase in diversity among the LV clade. Such forms first appeared in the Hauterivian and expanded tremendously over the next two stages. Note that if these two clades are excluded, diversity of LV and RV rudists remains very similar in the Barremian and Aptian. Comparing taxa at node 5, there is a statistically significant increase in the number of genera in the pachytragid group as compared with the other two, but this is not seen in phenon-level data (Fig. 4.5c&d).

Thus, although the evolution of an invaginated ligament may have been important, it was the reorganization of muscle attachments that correlates most closely with enhanced rates of morphological diversification in rudists. This case history demonstrates that paraphyletic groups allow only low-level generalizations to be made about life-habits, and fail to test at which level in the hierarchy significant evolutionary changes come into play. A proper understanding of morphological (and ecological) diversification can only come through better resolution studies in which grades are replaced by clades.

In summary, only monophyletic taxa have biological significance, which they derive from their unambiguous correspondence to the branching pattern of the evolutionary tree. Paraphyletic and polyphyletic taxa neither accurately reflect tree branching pattern nor have other than *ad hoc* correspondence to 'adaptive unity' or 'discrete areas of morphospace'. They are thus units selected from a large number of potential groupings through default or by subjective judgement based on convention.

Non-monophyletic taxa as samples of phena

Even if the majority of higher taxa represent non-monophyletic taxa, Sepkoski (1984, 1987) has argued that they still serve a role. Every appearance of a paraphyletic and polyphyletic taxon must correspond to the appearance of its first included phenon, and every last occurrence likewise corresponds to the last occurrence of its latest included phenon. Thus higher taxa, however constructed, provide a sample of phenon-level events. If traditional (i.e. non-cladistic) databases are dominated by non-monophyletic taxa, the great majority of higher taxa will represent 'random cuts' of the evolutionary tree and thus provide an unbiased sample of all phenon-level events. Higher taxa are therefore seen as a simple means of sampling phenon-level events.

The evolutionary patterns that are commonly deduced from the analysis of taxonomic distributions in the fossil record include studies of standing diversity, origination, extinction, and duration. However, monophyletic, paraphyletic, and polyphyletic taxa are not equally valid for such analyses because each has been created using different criteria. The implications of this are as follows.

Taxic diversity. The number of taxa present within specified time intervals provides a measure of the diversity of a clade. If a taxon is scored as present only in intervals in which one or more of its included phena are recorded, then it is immaterial which kind of higher taxon is used. All three will provide a method of sampling *observed phenon diversity* using non-overlapping sets.

However, if taxonomic ranges are interpolated (i.e. the taxon is assumed to have existed between its first and last occurrence in the fossil record, even though it may be missing at certain intervals within this range) then the situation differs. Range interpolation assumes that the first and last phenon in a taxon belong to the same unique branch of the evolutionary tree. Interpolation is valid for monophyletic taxa, which represent complete branches, and for paraphyletic taxa, which, although partial branches, maintain genealogical continuity between first and last members.

Interpolation is not valid, however, for polyphyletic taxa. The latter are constructed from phena belonging to two or more independent branches of the evolutionary tree, and therefore produce spurious results. If a polyphyletic taxon unites two or more monophyletic groups whose individual ranges are neither overlapping nor contiguous, then interpolation will imply that phena of one or other should be present during certain periods, when none in fact is. Alternatively, if a polyphyletic taxon comprises two or more overlapping monophyletic groups, interpolation will underestimate the number of evolutionary branches present in any interval.

Paraphyletic and monophyletic taxa are thus equally appropriate for measuring standing diversity when range interpolation is involved, but polyphyletic taxa are not.

Origination patterns. The first appearances of higher taxa through geological time are used as a proxy for first appearances of phena. Since the first record of monophyletic, paraphyletic, and polyphyletic taxa all correspond to the first appearance of a phenon in the fossil record, all three are equally appropriate for sampling observed origination patterns.

Extinction patterns. The last appearances of higher taxa through geological time are used as a proxy for phenon extinction patterns. The last appearance of a monophyletic taxon in the fossil record is defined by the last appearance of a phenon with the derived characters (synapomorphies) of that taxon. Since no later phenon is known to possess these characters, the disappearance of this phenon (and monophyletic taxon) corresponds to true biological extinction. It marks the termination of a branch of the evolutionary tree.

This is not necessarily the case, however, for paraphyletic taxa. These may disappear from the fossil record as a result of taxonomic convention, not biological extinction. The upper limit of a paraphyletic group is defined by the absence of a derived character or characters that demarcate other groups (see Fig. 4.1). In the simplest case a paraphyletic group will comprise a single lineage evolving through time. At a certain point, determined by a taxonomist, this taxon will 'go extinct' because the lineage acquires an additional character that has been used to define the start of another taxon. Note that the characters used to define the 'extinct' taxon throughout its range are still present in the descendant taxon – there has been no loss of genetic diversity. In reality the lineage continues even though taxonomic convention dictates that there is a change of nomenclature at this point. Disappearance for nomenclatural reasons associated with character acquisition is termed *pseudoextinction* (Van Valen, 1973a; Stanley, 1979; Patterson & Smith, 1987; Fortey, 1989; Valentine, 1990).

Many paraphyletic groups may include monophyletic branches within them. The last appearance of each of these represents true biological extinction. Therefore, if any of these monophyletic elements survive beyond the lineage that gives rise to the abstracted monophyletic group or groups, then the last occurrence of a paraphyletic group in the fossil record will mark a biological extinction event, not a pseudoextinction.

Last occurrences of paraphyletic groups thus may or may not record true biological extinctions. Each case needs to be examined individually, and this requires establishment of sister group relationships. As knowledge of phylogenetic relationships is needed to interpret paraphyletic groups correctly, such groups by themselves do not provide reliable data on extinction patterns.

Polyphyletic taxa are composites of two or more unrelated groups. These groups can be either monophyletic or paraphyletic. Thus, the last record of a polyphyletic taxon may often be appropriate for analysing patterns of biological extinction, but cannot be assumed to be so. Once again, phylogenetic analysis is necessary to clarify the nature of the groups.

Last occurrences of paraphyletic and polyphyletic taxa cannot be assumed to represent extinction events at phenon level, and it is therefore wrong to analyse extinctions using compilations of higher taxa whose phylogenetic status is unknown.

Taxonomic ranges and durations. These analyses use the geographic and/or stratigraphic distributions of taxa, as deduced by interpolation between end members, to define total range and duration of taxa. This information is then used to investigate patterns of taxonomic longevity or geographic distribution through time. The origin and extinction of monophyletic groups mark real biological events, and interpolation between end members is appropriate, therefore monophyletic taxa have taxonomic ranges and durations that are meaningful. Polyphyletic taxa are artificial amalgams of two or more distantly related groups, and therefore exist only in the minds of taxonomists and arise by mistake; their range and duration have no biological significance.

Paraphyletic taxa have an origination and their last occurrence may or may not represent a biological extinction. If the last occurrence is a pseudoextinction, the taxon's range is truncated through taxonomic convention and the duration of the taxon has no significance. If the last occurrence represents an extinction, there is still a problem because the origination of the paraphyletic taxon is not necessarily the origin of the branch that is recognized as becoming extinct. The correct origin of the subset of taxa whose extinction has been identified must often lie within the paraphyletic group, leading to an overestimation of the true range. Thus, paraphyletic taxa are inappropriate for such studies.

In summary, only monophyletic taxa have a birth, history, and death in biological terms. They are therefore the only appropriate kind of taxon to use in the analysis of evolutionary patterns. However, patterns of origination can be established using both paraphyletic and polyphyletic taxa by assuming that higher taxa provide a random sample of phenon-level events. Standing phenon-level diversity can also be estimated using paraphyletic, but not polyphyletic taxa. However, it is worth remembering that the random sample of phena derived from a predominantly non-monophyletic taxonomic data base will only provide an estimate of observed (sampled) phenon diversity. Since we know that 'sampling effects exert a control on total apparent species that is so profound that it may be impossible to determine trends in actual worldwide species richness from trends in the total number of described species' (Signor, 1978, p. 405), observed phenon diversity will differ from real species diversity in that it is strongly affected by the quality of the fossil record (Chapter 5). For this reason a phylogenetic approach, which is much more able to compensate for sampling deficits, is preferable.

Classification and rank

So far we have only been concerned with the problem of how to group phena into higher taxa. It is now appropriate to consider how these groups are named and classified. Our nomenclatural system originates in the work of Carl von Linné in the 1750s, although various hierarchical schemes had been proposed before then (see Nelson & Platnick, 1981 for an excellent review).

The Linnaean scheme is hierarchical in structure: phena are grouped into taxa and these taxa themselves are grouped into progressively more inclusive taxa (higher taxa). In this way a classification comprises a nested set of taxa, each more inclusive than the last. Each level in the hierarchy is given a formal name, the most widely used forming the sequence: kingdom, phylum, class, order, family, genus, and species.

Classifications of historical groups are inherently hierarchical (Wiley, 1981). The utility and ubiquitous adoption of the Linnaean scheme of classification arose because of its structural correspondence with the hierarchy of characters in organisms. As the homologies distinguishing taxa are hierarchically arranged, it is possible to construct a classification that matches the pattern of inferred phylogenetic relationships precisely.

The primary purpose of a classification is to group together phena in the most informative way possible. Phena are usually grouped on the basis of shared characters. The most informative classification is thus the one that provides a summary of the maximum number of attributes of the phena included. A classification encodes the maximum amount of information when it bears a one-to-one correlation with the most parsimonious cladogram of relationships (Farris, 1977, 1979, 1980, 1982a). Thus, classifications ought to begin from a cladogram of inferred relationships.

Attributes of rank

The formal rank bestowed on a taxon is indicative of its relative level of inclusiveness only (Hennig, 1966). To say that a group is a family, for example, implies only that it is included within higher hierarchical levels (orders, classes, etc.) and that its members may themselves be included within lower hierarchical levels (genera, phena). Unfortunately, rank has been applied to groups for a variety of reasons and is often mistakenly taken to imply other attributes.

Smith (1988a) identified the following five reasons why a high categorial rank had been applied in traditional classifications:

1 *High categorial rank as a topological consequence of a group achieving considerable diversity.* When a taxon includes a large number of phena it is given a high taxonomic rank simply because of the hierarchical structure of the Linnaean system. Groups that achieve high diversity require a large number of Linnaean categories to classify them effectively and thus end up with a high categorial rank. This corresponds to the original intention of the Linnaean classificatory system; for example, all five extant classes of echinoderms have achieved high rank for this reason. Note that high rank here is simply a convention of classification and carries no implication about the amount of morphological disparity involved at the time of origination.

2 *High categorial rank as a consequence of perceived morphological distinctiveness.* Some workers (e.g. Paul, 1979; Sprinkle, 1983; Campbell & Marshall, 1986) have argued that taxa should be given high categorial rank on the basis of morphological distinctiveness, irrespective of their size (number of included phena) or longevity. This morphological distinctiveness may be: (i) real; (ii) the result of emphasis on one or two distinctive apomorphies; or (iii) simply the result of misinterpretation. For example, a new class of echinoderm, Concentricycloidea, was created recently (Baker *et al.*, 1986) for a single phenon of Recent deep-water echinoderm. This class was erected be-

cause of the derived morphological difference in ambulacral structure between *Xyloplax* (the concentricycloid genus) and other starfishes. When sister group relationships were analysed, *Xyloplax* was shown to be most closely related to another deep-water asteroid family, Caymanostellidae, and nested deep within the Asteroidea (Smith, 1988b; Belyaev, 1990).

3 *High categorial rank as a result of sister-group relationship.* High taxonomic rank is often applied to a group which is neither perceived to have great morphological distinctiveness, nor has subsequently achieved great diversity, but which is sister-group to a highly diverse taxon. The practice of applying equal rank to sister groups, if taken to its ultimate logic, can mean that a sequence of fossil phena along a stem lineage are each given successively higher rank. More usually, however, such groups are created when a fossil is discovered that is sister group to a well established clade. If this has some but not all of the traditionally recognized features of the well established clade then it is often split off as a new taxon of equal rank to the established clade. An example is the Burgess shale genus *Echmatocrinus*, which was erected as a new subclass of crinoid (Sprinkle & Moore, 1978) because it had uniserial arms but lacked the well-defined theca that characterized all other crinoids.

4 *High taxonomic rank is given to a paraphyletic ancestral group after abstraction of a number of well defined monophyletic groups.* Here high taxonomic rank is given not for morphological distinctiveness, but for plesiomorphy and the assumption that the group has given rise to a number of taxa recognized at high categorial rank themselves. The 'phylum' Procoelomata, created by Bergstrom (1989) for Macheridian-grade animals comprising the stem group from which most other Metazoan phyla are derived directly, is a splendid example of this.

5 *High categorial rank has been given because of ignorance.* In a few cases monotypic taxa that are improperly understood, or have been misinterpreted because of poor or incomplete knowledge, have been elevated to high taxonomic rank simply because the systematist has not been able to place them. With later work, such taxa are reclassified into existing schemes.

Taxonomic rank in traditional (non-cladistic) classifications has become hopelessly muddled because taxonomists may have bestowed rank for several disparate reasons. Lack of consistency renders rank largely meaningless, although some workers implicitly or explicitly continue to believe that rank in some way correlates with morphological disparity. Hennig (1981) argued that Linnaean ranks should be abandoned in favour of a numerical convention to represent set membership. Some later workers (e.g. Ax, 1987; Craske & Jefferies, 1989; de Queiroz & Gauthier, 1992) have continued to call for the complete abandonment of Linnaean classification schemes. This is because the very existence of categorial ranks 'encourages spurious comparisons between entities assigned to the same rank but that are not otherwise comparable' (de Queiroz & Donoghue 1988, p. 334; see also Gauthier *et al.*, 1988).

However, most cladists have continued to use a Linnaean hierarchical classification, with rank used solely to indicate relative inclusiveness.

Cladistic classifications and the problem of fossils

Cladistic classifications are those that recognize only monophyletic taxa. As in all classifications, some form of ranking convention is necessary to place taxa into their appropriate level of inclusiveness within the tree of life. De Queiroz & Gauthier (1992) argued that simply indenting taxa on the printed page suffices to establish rank. However, systematists need to communicate and it is useful to know at approximately what level of inclusiveness a taxon has been placed without having to be totally familiar with the details of individual nomenclatural schemes. Since the Linnaean scheme is universally recognized and understood, it makes sense to retain this as a means of indicating relative inclusiveness. The particular level at which a rank is applied remains the selection of a taxonomist, but a few simple rules ensure that rank is relatively consistently applied within a single hierarchical scheme (Wiley, 1979, 1981; Schoch, 1986).

Fossils pose a special problem in cladistic classifications because groups with an extensive fossil record include large numbers of low-diversity, extinct clades as progressively more distantly related sister groups to each of the modern clades. If rank were to be assigned on the basis of inclusiveness, each successive extinct sister group would have to be given an ever more inclusive taxonomic rank. For some groups, e.g. brachiopods or crinoids, this would enormously inflate the number of taxonomic ranks required, yet would serve little useful purpose.

It is clear that fossils should not be treated as different from extant organisms when establishing the phylogenetic relationships of a group (see Chapter 3). To do so could result in a mistaken phylogenetic hypothesis. However, purely practical aspects dictate that fossils ought to be treated differently when it comes to transforming a phylogenetic hypothesis of relationships (cladogram) into a classification. Here the distinction between crown groups and stem groups is important.

Although Hennig (1969) first identified the distinction between crown groups and stem groups, it was Jefferies (1979) who formalized the names and established their importance. A *crown group* is an extant monophyletic taxon that also includes extinct phena with all the diagnostic attributes of that taxon (or can be shown to have secondarily lost such attributes). In evolutionary terms it is composed of the latest common ancestor of the extant members of the taxon and all of its descendants (Figs 4.6 & 4.7). In contrast, a *total group* is an

extant monophyletic taxon that contains all extinct phena possessing
one or more of the diagnostic attributes of the taxon. In evolutionary
terms it includes every phenon that is more closely related to the
modern representatives of the taxon than to its nearest living sister
group (Fig. 4.6). The *stem group* is the total group with the crown
group abstracted, and is thus paraphyletic (Figs 4.6 & 4.7).

The distinction between crown groups and stem groups is important
because it differentiates between fossil taxa that are subsumed within
an existing taxon (fossils within the crown group), and those that
would simply inflate the taxonomic hierarchy needlessly (members of
the stem group). Since all fossils can be placed in a stem group at some
level of the hierarchy, the task of palaeontological systematists is to
place fossils in their correct stem groups and establish sister group
relationships among stem group taxa.

In an analysis that includes both fossil and Recent members, the
resultant cladogram will probably consist of an interlaced series of
extant and extinct taxa. The stem group fossils are those that lie

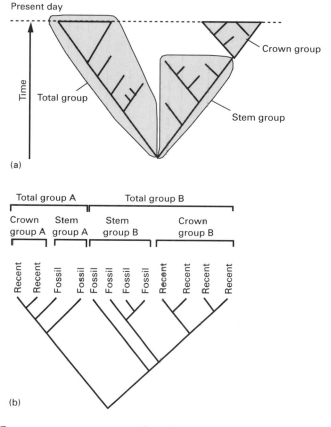

Fig. 4.6 Crown group, stem group, and total group concepts. (a) Hypothetical
evolutionary tree. (b) Cladogram showing how these groups are identified in
practice.

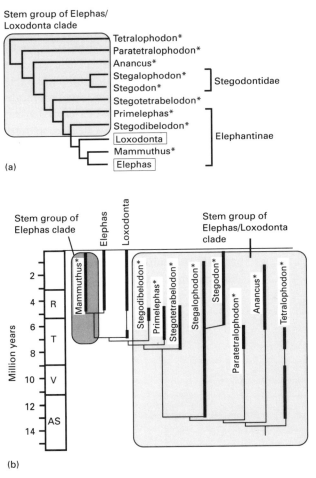

Fig. 4.7 (a) Cladogram and (b) evolutionary tree for higher pachyderms (Tassy, 1991).
Stem groups are identified by shaded boxes. *, Extinct.

between the node separating two extant sister group taxa and the first
dichotomy leading to extant phena that occurs within one of the sister
groups (Fig. 4.6b).

Patterson & Rosen (1977) proposed that all extinct monophyletic
taxa constituted *plesions* which could be inserted anywhere within the
classification without altering the Linnaean rank of the crown group.
Plesions might comprise just a single phenon, or could encompass a
vastly diverse group such as the Ammonoidea. The important point
about plesions is that they are classified from the termini to the basal
node, and their categorial rank is determined according to the number
of subordinate ranks that are required to accommodate the included
phena. Thus, the primary Linnaean hierarchical framework is tied to
the nested relationships of extant phena alone (though these should
have been established on the basis of all available data, using both
extant and extinct phena).

Crown and total groups could be given separate names, as recommended by de Queiroz & Gauthier (1992), but this would simply expand the classification scheme to no advantage. Only the total group need be given a formal name and Linnaean rank. Thus, the name and rank given to the clade formed by the modern phena is extended to include all stem group members of that taxon. Plesions are then listed before the crown group in order of their appearance on the cladogram. When a plesion is included in a classification it is usual to give its assigned rank in brackets immediately following (Table 4.2).

Table 4.2 Linnaean classification scheme for arbacioid echinoids based on the cladistic analysis discussed in Chapter 6 and shown in Fig. 6.7. Taxa are ranked according to their cladogenic order unless stated otherwise by the tag *sedis mutabilis*. An asterisk* after the name indicates that the group is plesiomorphic at the level of analysis so far achieved and requires further investigation to establish its monophyletic elements

Order Arbacioida Gregory
Plesion (Genus) *Hypodiadema* Desor*
Plesion (Family) Hemicidaridae Wright
    Subfamily Hemicidarinae Smith & Wright
      Genus *Hemicidaris* Agassiz*
      Genus *Hemitiaris* Pomel
      Genus *Plesiocidaris* Pomel
    Subfamily Pseudocidarinae Smith & Wright
      Hemicidaris termieri Lambert*
      Genus *Pseudocidaris* Etallon
      Genus *Cidaropsis* Cotteau*
Plesion (Genus) *Gymnocidaris* Agassiz*
Unnamed plesion 1
  Family Acropeltidae Lambert
    Genus *Acropeltis* Agassiz
    Genus *Goniopygus* Agassiz
  Family Glyphopneustidae Smith & Wright
    Genus *Glyphgopneustes* Pomel
    Genus *Arbia* Cooke
Unnamed plesion 2
    Genus *Glypticus* Agassiz
    Genus *Asterocidaris* Cotteau*
  Family Arbaciidae Gray
    Unnamed subfamily 1
      Plesion (Genus) *Codiopsis* Agassiz*
      Genus *Dialithocidaris* Agassiz
      Genus *Pygmaeocidaris* Doderlein
    Unnamed Subfamily 2
      Group 1
      Genus *Arbacia* Gray
      Genus *Tetrapygus* Agassiz
      Group 2
      Genus *Coelopleurus* Agassiz
      Genus *Murravechinus* Philip*

Note that, according to this convention, rank is determined independently for crown groups and plesions. In the hierarchy tied to extant taxa, sister groups are given equal rank and thus a high rank can still be given to a taxon with very low diversity if its sister group has high diversity. Therefore rank serves primarily to group phena on their relative time of divergence and is topologically controlled. However, in the case of plesions, rank is based solely on subordination and is dictated by the diversity of phena it contains. There is no correlation between the rank of a plesion and its time of appearance.

Thus, although fossils have to be treated in a slightly different way when it comes to constructing a classification, the use of plesions alleviates many of the problems. This means of course that taxonomic rank does not have a single consistent basis for its definition, but it never has had. Provided rank is used solely to denote relative inclusiveness and not to deduce, for example, morphological disparity or equivalency among taxa, then the system works very well.

Macroevolution and emergent characters of higher taxa

Macroevolutionary processes: essential requirements

Whereas microevolution is concerned with the study of heritable variation at the level of the population, macroevolution embraces evolution at and above the species level (Salthe, 1975; Stanley, 1979; Eldredge & Cracraft, 1980; Eldredge & Salthe, 1984; Vrba & Eldredge, 1984; Padian, 1989; Valentine, 1990). The differences between micro- and macroevolution are thus in the level of hierarchy. Some workers, most recently Levinton (1988) and Hoffman (1989), have forcefully argued that macroevolution is simply the extrapolation of observed microevolutionary processes into geological time. Others have argued that it is more than simple microevolution and that taxa have properties on which selection can work that are simply not present at lower levels. Some who support macroevolution as a discrete phenomenon (e.g. Eldredge & Salthe, 1984; Ghiselin, 1984b) restrict it to the species level, while others permit taxa of all rank to participate (e.g. Valentine, 1990).

There are two distinct processes grouped under the term macroevolution: sorting and selection (Vrba & Eldredge, 1984; Cracraft, 1982, 1989). Sorting refers to the differential survivorship of taxa within clades as perceived in the fossil record. Selection is a process whereby the relative survivorship of a taxon can be causally linked to the possession of a heritable property not found at lower levels in the hierarchy.

For taxa to participate in selection they must fulfil three criteria: (i) they must correspond to real biological entities; (ii) they must have heritable properties that are not found at lower levels in the hierarchy ('emergent properties', Salthe, 1979); and (iii) they must have the potential to give rise to other taxa so that emergent properties can be selected through differential survival.

Taxa as real biological entities. As discussed above, only a hierarchy of monophyletic taxa exists independently of taxonomists' convention (Patterson, 1982; Schoch, 1986). Monophyletic taxa have an origin, a history, and ultimately a biological extinction. Paraphyletic taxa have an origin but only a partial history, since taxonomic practice has removed one or more derived portions of the group. They cannot therefore be considered as 'individuals' or independent of taxonomic convention. Polyphyletic taxa have no biological reality and arise through taxonomist error. Thus, only monophyletic taxa can be considered as players in macroevolutionary phenomena.

Species are often thought of as 'individuals' immune to the problems of paraphyly (e.g. Eldredge & Cracraft, 1980). This has led to the view that species are distinct from taxa. In Chapter 2 I argued that species, as defined from a genetic stance, are indeed distinct from taxa, but that palaeontological 'species' (phena) are for the most part taxa. As such, they are just as prone to the problems of paraphyly and polyphyly as any other taxon. They are simply taxa lying at the limit of resolution afforded by available data.

Emergent properties of higher taxa. A property that exists at higher levels in a hierarchy, but which logically cannot exist at lower levels, is an emergent property of the hierarchical level at which it first appears (Salthe, 1979). Thus there is an important distinction between *group properties* and *emergent properties* at any hierarchical level (Vrba & Eldredge, 1984). Group properties are properties that happen to be present in every unit in a taxon rather than a property unique to the hierarchical level. For example, all the phena in a clade might have the morphological characteristics that are diagnostic for that particular clade, but these are group characteristics that are found at all levels from the individual upwards. Emergent properties are traits that can only exist above a certain level in the hierarchy, and their existence is debatable at best (Cracraft, 1989; Hoffman, 1989). The most commonly cited emergent properties are those determined by the demographic structure of taxa (Vrba & Eldredge, 1984). However, Cracraft (1989) has pointed out that demographic properties are determined to a large extent by the external environment and are not necessarily intrinsic to the taxon.

In fact it is hard to see that any higher taxon can have biologically predetermined emergent properties in the same way that longevity or fecundity in individual organisms is predetermined. Taxa are bounded by extrinsic, historical factors and can last indefinitely, whereas organisms have finite lifespans and must reproduce to survive. Thus, Vrba & Eldredge (1984, p. 155) concluded, 'it is problematic whether any emergent species characters interact with the environment to causally influence speciation rate'.

Some emergent properties are postulated because of their discordant behaviour across hierarchical levels. Holman (1989) compiled taxonomic data on duration, origination rate, extinction rate, and proportion of originations for phyla, orders, and classes and found a 'significant' difference between the pattern displayed by phyla and that displayed by classes and orders. This discordance was used to argue that there was a 'real' difference between phyla and classes.

However, discordant patterns may arise purely for topological reasons. The geometry of evolutionary trees dictates that major groups tend to arise early in the history of a clade (Raup, 1983). Since the Linnaean system of nomenclature is also hierarchical in structure, and high taxonomic rank has traditionally been given to groups that achieve high diversity, it is also an inevitable consequence of tree topology that higher taxa in general appear earlier in the fossil record than taxa lower down in the hierarchy. The observation that phyla reach a peak in diversity during the Phanerozoic earlier than classes, which themselves peak before families, was interpreted as biologically significant by Valentine (1969). Yet the pattern can be explained in topological terms and needs no biological explanation. Topological attributes of rank have here been mistaken for biological attributes.

Jablonski & Bottjer (1990b) noted a difference in environment of origin along an onshore–offshore gradient between orders and families of marine invertebrates. Whereas orders appear to originate preferentially in onshore settings, no such pattern was discernible at generic level. This, they argued, differentiates taxa of high rank from lower-level taxa. As demonstrated below, this also arises from mistaking topological constraints of the hierarchical system for biologically significant traits.

Heritability of attributes. For selection to occur, emergent properties must be heritable, such that differential sorting can act through time. Even if higher taxa have emergent properties, they can only participate in macroevolutionary phenomena if they give rise to other higher taxa which inherit those emergent properties. Supra-specific taxa are not generally considered able to give rise to other taxa: families do not give rise to other families, it is species that give rise to families (Eldredge & Salthe, 1984). Nevertheless, Valentine (1990) has argued that this is a result of recognizing only monophyletic taxa. He believed that paraphyletic taxa could be considered as parental to other taxa and thus able to participate in macroevolutionary phenomena.

In fact it is even very debatable whether species can be considered to give rise to other species. Species do not speciate, they are the products of speciation (Cracraft, 1989). It is reproductively cohesive populations within existing species that give rise to new species through divergent evolution. Species as taxa in the fossil record originate and go extinct, but if they give rise to other species then they must be paraphyletic and thus represent arbitrary constructs (Nelson, 1989b).

Here we are faced with a contradiction. Only monophyletic taxa have any claim to represent biologically meaningful units of macro-evolution, but by their very definition they are the products of specia-tion, not active participants. Yet, only by giving rise to other taxa can the process of selection come into play through differential survival of the progeny. Phena, the 'species' of the fossil record, comprise a mix-ture of three kinds of entities (see Chapter 2):

1 Populations from a single locality and horizon (presumably inter-breeding populations).

2 Groups of populations sharing a derived character state, i.e. mono-phyletic taxa, which differ in no way from higher monophyletic taxa (Mischler & Donoghue, 1982; Nelson 1989b).

3 Groups of populations united on plesiomorphy alone, i.e. metataxa, which may or may not be paraphyletic (de Queiroz & Donoghue, 1988, 1990a,b).

It is simply wrong to treat all phena as 'individuals' and thus players in macroevolutionary scenarios, as Stanley (1979) and others have done. The concept of 'individuality' has been used in various senses, and 'simply asserting that something is an individual . . . does little to clarify the nature of its existence' (de Queiroz & Donoghue, 1988, p. 322). Phena have no claim to biological integrity in the way that reproductively cohesive populations do, and the three categories listed above find commonality only in that they are the smallest identifiable unit recognizable to a taxonomist.

Macroevolutionary models built up from analysis of the stratigraphic ranges of phena are just as invalid as are those derived from higher taxa, and for the same reasons. Any phenon recorded from multiple stratigraphic levels that gives rise to another phenon must be paraphy-letic and thus artefactual both in theory and in practice (Nelson, 1989b, p. 74) and no artefactual grouping can have emergent properties other than in the taxonomist's mind. Only single populations satisfy the criteria of being real biological entities that give rise to other like groups. Whether these have true emergent properties is a moot point, but anyway they are by definition participants in microevolution, not macroevolution. I can only conclude therefore, like Eldredge (1986, p. 358) and Hoffman (1989), that macroevolution – in the sense of natural selection – cannot occur above the level of the individual organism.

If species selection appears groundless, what about *species sorting*, the differential perpetuation of species lineages due to upward causa-tion (Vrba & Gould, 1986)? This derives its empirical basis from the observed stratigraphic ranges of phena. The sorting of species over time, through differential speciation and extinction, leads to the waxing or waning of clades relative to one another. Sorting can arise through natural selection at the level of individual organisms, species selection (which is here rejected), or a variety of non-selective mechanisms that influence rates. But, here again, if phena do not have intrinsically determined emergent properties, it is hard to see how sorting of phena is any more than the summation of sorting at lower levels in the hierarchy and thus an epiphenomenon with no explanatory powers.

A case study: onshore origination of higher taxa

One of the strongest cases so far put forward in support of a hierarchical view of evolutionary processes and the reality of emergent properties comes from the work of Jablonski & Bottjer (1990a,b, 1991). They compiled detailed documentation of the environmental distribution of first appearances of orders of benthic marine organisms in terms of five onshore–offshore environments. Environments of first occurrence were inferred from the sedimentary and stratigraphic features of the rock containing the oldest known phenon of each order, and ranged from 'nearshore' to 'slope and basin'.

Jablonski & Bottjer found a strong bias towards onshore origination for orders of taxa with a good fossilization potential. On the other hand, taxa with a poor fossilization potential demonstrated no preferred onshore–offshore pattern. Furthermore, generic level originations were found to be diversity dependent: the number of new taxa arising in each environmental setting was proportional to the overall taxonomic diversity of the clade across those environments (Jablonski & Smith, 1990; Jablonski & Bottjer, 1991).

From these findings Jablonski & Bottjer (1991) deduced that the pattern of ordinal originations could not be predicted from the pattern of lower taxonomic originations and argued that this lent strong support to the hierarchical view of emergent characters. It suggested to them that there was a biologically significant difference between groups of high and low taxonomic rank. Consequently they proposed that the 'origin of new body plans, with the potential to diversify and accumulate additional derived characters, is governed by factors different from those that determine origination of species that simply produce more species or new genera' (Jablonski & Bottjer, 1991, p. 1832).

Since this study has wide implications and is the best documented evidence for the existence of discordance between taxonomic levels in a feature of evolutionary significance, it is worth considering the available evidence in detail.

The onshore–offshore hypothesis is particularly sensitive to sampling biases, since Jablonski & Bottjer's approach relies on averaging over the entire 250 Ma of the post-Palaeozoic. There should therefore be no temporal bias in either the development of continental shelf area or in taxonomic originations. Unfortunately, both vary over time.

Temporal bias in marine continental shelf area. The relative development of onshore and offshore facies varies temporally with global eustatic changes in sea-level. This is clearly seen from the sea-level curve of Haq *et al.* (1987), where there is a strong onshore bias from the base of the Triassic to around 150 Ma, and a strong offshore bias from 100 Ma to about 35 Ma (Fig. 4.7). Furthermore, because the sections in western Europe and North America are so much better studied than those from elsewhere, there is also a very strong bias in the geographic locations of first occurrences. Thirty-three of the 49 records listed by Jablonski & Bottjer (1990b) are from western Europe, predominantly France and England. The onshore–offshore distribution of facies will therefore be strongly tied to the history of the western European margin, and in particular to sea-level changes associated with the opening of the North Atlantic.

Temporal bias in the origination of higher taxa. The distribution of taxonomic originations is also not uniformly distributed through time. This is certainly the case in Fig. 4.7, where 15 out of 25 of the ordinal taxa regarded as having good preservation potential arise in the first 60 Ma period (<24% of the total time), and only four appear in the last 100 Ma. This contrasts with the distribution of ordinal taxa deemed to have poor preservation potential, where nine appear in the first half and seven in the second half. It also contrasts with the distribution of generic originations; for example, for crinoid genera cited by Jablonski & Bottjer (1990b) 10 predate and 11 post-date 125 Ma (which can be taken as midway, in terms of both time and of the shift from predominantly onshore to offshore sea-level curve). Similarly, for genera of the echinoid order Calycina, eight predate and 12 post-date 125 Ma.

This shift in general lithofacies distribution from predominantly onshore during the Triassic and Jurassic to predominantly offshore during the mid-Cretaceous is confirmed when the distribution of individual fossil localities for Calycina is examined (Jablonski & Smith, 1990, unpublished data). For records that can be placed in their onshore–offshore environmental setting, there is a shift in the offshore: onshore ratio from approximately 1:4 for the interval 100–200 Ma, to 2:1 from 100 to 0 Ma. The change in ratio is most marked at around 125 Ma. However, averaged over this entire period the onshore:offshore ratio for all sites is close to 1:1. If we look at environmental settings for ordinal originations in three periods defined by the relative extent of eustatic onlap (Fig. 4.8), there is a clear correspondence. Onshore originations predominate when offlap predominates in the sea-level curve.

Do orders therefore show a pattern of onshore origination that could not be predicted from the distribution of genera? Orders appear earlier in the fossil record primarily because of topological properties of the Linnaean hierarchical system. It is simply an inevitable outcome of branching topology that higher taxa tend to appear earlier in the fossil

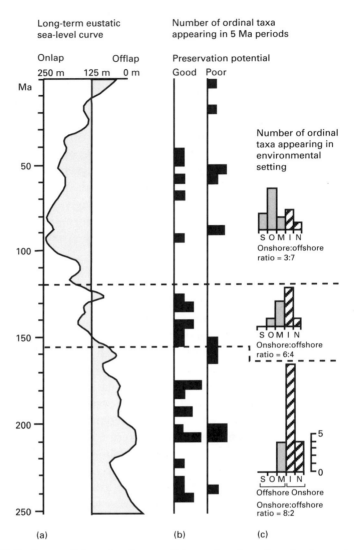

Fig. 4.8 Onshore–offshore data for post-Palaeozoic orders of marine invertebrates (Jablonski & Bottjer, 1991). (a) The long-term eustatic sea-level curve of Haq *et al.* (1987) for the last 250 Ma. The median onlap is about 125 m and this highlights the change in onlap/offlap that took place around 120–150 Ma. (b) The numbers of higher taxonomic originations (orders) appearing through time, separated into those with a high preservation potential and those assessed to have a low preservation potential (after Jablonski & Bottjer). (c) The onshore : offshore ratio of taxonomic originations for the three periods indicated shows a marked shift from predominantly onshore to predominantly offshore correlated to the eustatic sea-level curve. S, slope and basin; O, outer shelf; M, middle shelf; I, inner shelf; N, nearshore. S+O+M = offshore facies; N+I = onshore facies.

record than the majority of lower taxa, e.g. genera. Since there is also a temporal bias in the development of offshore–onshore facies, it is important to compare patterns of first appearances of orders and genera over the same time interval. Of the 25 orders deemed to have good fossilization potential, 15 are echinoids, of which 13 appear prior to 125 Ma. The ratio of onshore:offshore originations for these 13 taxa is 3:1. If we take the originations of echinoid genera within the order Calycina (Jablonski & Smith, 1990) for the same period, we have 11 originations, with an onshore:offshore ratio of, again, 3:1.

Thus there is no difference in the environmental setting of first occurrences of phena classified as orders and those classified as genera once sampling biases have been removed, and the hypothesis that orders have 'emergent biological properties' in this case has arisen through mistake. Neither the distribution of taxa of different rank, nor the relative development of onshore–offshore facies on the continental shelf of western Europe, are randomly distributed through time; simple topological attributes of a classificatory scheme have led to erroneous biological deductions.

Summary

Because of the vagaries of the fossil record, phena are usually assembled into more inclusive taxa to document evolutionary patterns. The way in which this is done is important because it determines whether the groups have objective or conventional boundaries. There is no argument as to the reality of monophyletic taxa, nor about the artificiality of polyphyletic taxa, but the status and utility of paraphyletic taxa is disputed. It is argued here that paraphyletic groups serve only to mislead in the analysis of evolutionary patterns. The view that paraphyletic groups represent useful clusters of 'adaptively unified' phena or discrete clusters in morphospace is more a statement of faith than observation, since most are created by default. Alternatively, higher taxa can be thought of as simply sampling phenon-level events. However, the problem then arises that a random sample of phenon-level events will simply mirror the quality of the fossil record through time and need not identify meaningful biological patterns. Consequently, only monophyletic taxa should be used in the definition of evolutionary patterns.

The Linnaean rank applied to a taxon is a matter of convention, but remains useful for communicating information on a taxon's relative level of inclusiveness. However, there is no justification for treating taxa of equivalent rank across disparate clades as in any way commensurate. Only sister taxa represent truly comparable taxa.

Fossils are best included into a classification by application of the plesion concept. This ensures that taxonomic rank and nomenclature is not inflated unnecessarily by large numbers of extinct stem-group taxa. The taxonomic rank of each plesion is derived independently whereas rank among crown groups forms a single integrated reference system.

Finally, the concept of macroevolution is explored from a cladistic viewpoint and is rejected. For species selection to be viable, species must be: (i) real biological entities; (ii) have emergent properties at the appropriate taxonomic level; and (iii) give rise to other taxa of the same rank. Only monophyletic taxa are real biological entities and any taxon that gives rise to another must be paraphyletic and hence artificial. No monophyletic taxon gives rise to another taxon of the same rank; even species do not give rise to species, but arise from differentiation at the population level. Finally, emergent properties remain unproven. Emergent properties are usually inferred from observed discordant behaviour of taxa at different levels of the hierarchy. But this by itself is not enough. The suggestion that higher taxa preferentially originate in onshore settings arises because of the Linnaean convention of ranking, and not as a result of any biological phenomenon involving emergent properties.

5 The nature of biostratigraphic data

Biostratigraphy is concerned with the construction of a relative time scale through the study of species distribution in the geological record. It is based on the observations that different strata contain different suites of phena and that, for all intents and purposes, the succession of these assemblages is non-random. These patterns of spatial order of appearance and disappearance of phena are referred to as *homotaxial patterns* (e.g. Harper, 1980; see also Schoch, 1986).

Homotaxial patterns exist independently of phylogenetic hypotheses. Any group has a definable stratigraphic range, provided it is consistently recognized by systematists. Monophyletic, paraphyletic, and even polyphyletic groups all have a first and last record in the geological column: it is only the biological significance that can be placed on these events that differs. Thus, dinosaurs (a paraphyletic group) predate passerine birds (a monophyletic group) in the geological record, irrespective of whether birds arose from among the dinosaurs. This is important because phena comprise both monophyletic clades and potentially paraphyletic metataxa (Chapter 2).

What is crucial, however, is that all such groups should be based upon clear and consistent morphological criteria, and must be consistently recognized if they are to be of use. Uncritical use of literature-based records to compile species range charts can result in very different patterns to those derived from an equally extensive, but taxonomically consistent, data set generated from study of the actual specimens (Culver *et al.*, 1987). Thus biostratigraphy must start from a sound basis of alpha taxonomy (i.e. the identification and description of phena).

Within single sections the raw biostratigraphic pattern is deduced simply by recording the distribution of phena against a measured lithological section. To obtain total ranges it is of course necessary to interpolate between the first and last observed occurrence for phena in the section. Thus, except for use in highly local correlation, gaps in the observed range of phena are discounted.

However, this tells us only about the total observed ranges for phena in that one section. There are very many reasons why local ranges may depart significantly from the total geological range of a taxon, e.g. sampling and preservational failure, local facies variation, and migration and local extinction of faunal or floral elements.

Biases affecting taxonomic ranges

In the more analytical of the biological sciences, sampling strategy needs to be planned before data are collected so as to avoid or minimize biases that might affect subsequent analysis. The major problem for analytical palaeontology is that the fossil record represents an already highly biased sample of past life. If sampling biases inherent in the data are not recognized or redressed in some way, there is a danger that observed patterns will be ascribed biological significance when in fact there is none. Sampling biases can affect origination, extinction, and taxonomic durations to create an apparent decline in per-taxon rates through time even when none exists (Pease, 1985, 1988a,b,c,d, 1992). For example, Thackeray (1990) observed that both background and mass extinction decline in intensity (as measured by percentage extinction rate) towards the present, and he sought a biological explanation for this. However, Pease (1992) demonstrated that a decline in origination and extinction rate is expected as a result of better sampling, more rock outcrop, and a decline in stage length towards the Recent (Fig. 5.1). Thus the task confronting palaeontologists is to discriminate between those patterns in the fossil record that represent real biological phenomena and those that simply represent sampling biases. This is far from simple, as the studies of Pease have shown.

Biases that affect the fossil record fall into two broad categories, those that are intrinsic to the species themselves and those that result from extrinsic geological processes. Inaccurate ranges that result from mistaken taxonomy pose a separate and often significant problem (Culver *et al.*, 1987; Smith & Patterson, 1988; Benton, 1989; Maxwell & Benton, 1990; Sepkoski, 1993) but are not considered here.

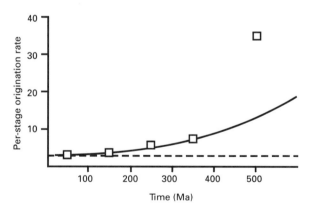

Fig. 5.1 The effects of sampling bias – estimated per-stage origination rates for bivalve families (Pease, 1992). The solid line shows the slope of per-stage origination rates predicted to be observed when sampling biases are taken into account (decreasing rock outcrop with increasing age and increasing average stage length) when origination rate is constant (dashed line). The per-stage origination rate that one would expect to observe declines towards the present even though origination rate is constant. The squares mark observed rates using data on bivalve families.

Factors intrinsic to taxa

Relative abundance. The relative abundance of a taxon directly affects the chances of it being preserved, and, if preserved, of it being found. Thus a species that occurs in relatively low population densities is less likely to be discovered and recorded as a fossil than a species that lives in high densities in the same environment. Phenon abundance typically follows a hollow-curve distribution, with a few taxa much more abundant than the rest (Willis, 1922, 1940; Dial & Marzluff, 1989). As most phena occur relatively infrequently in the fossil record (Buzas *et al.*, 1982; Koch, 1987; Smith & Patterson, 1988; Koch, 1991) sampling can have a profound effect on the perceived range of taxa. The effect of sampling density is clearly shown, for example, by data from Surlyk (1982), where numbers of specimens and numbers of phena of brachiopods from a late Cretaceous section are recorded (Fig. 5.2). The two plots match very closely, suggesting that observed phenon diversity is largely controlled by the sample size available at each interval.

The crucial effect that sample size can play in moulding taxic pattern has been clearly demonstrated by Koch (1987), who found that even large and equal sample sizes (1000–10 000 recorded occurrences) drawn from the same material will show differences of 8–25% in taxic composition (Fig. 5.3). When the two samples are unequal in size, the percentage of unique phena found in the larger sample increases in proportion. These differences become much more acute as sample size decreases. So, for example, comparing a set of 1000 occurrences with one of 125 occurrences *drawn from the same sample*, it is predicted that 70% of phena found in the larger data set would not be found in the smaller. Thus, sampling can obviously have profound effects on the observed homotaxial pattern. The ranges of relatively rare phena are likely to be significantly underestimated, all other things being equal.

Palaeontologists often want to compare faunas on either side of a boundary, and this is done by counting disappearances and originations across the boundary. Yet, even if the faunas on either side are identical, unequal sampling can generate spurious patterns. If more material is studied below the boundary than above, then sampling alone will generate more 'extinctions' than 'originations' at this boundary. Conversely, if more material is available from above the boundary, the reverse is true.

Even the type of sediment can have a possible effect on sampling and thus on observed species diversity and duration. Sediments that are only loosely consolidated and from which fossils are easily collected are likely to be searched more extensively than a similarly fossiliferous, but highly indurated or well-cemented sediment. The result is apparently higher faunal diversities in more easily collected lithofacies. There is even a correlation between the recorded taxonomic

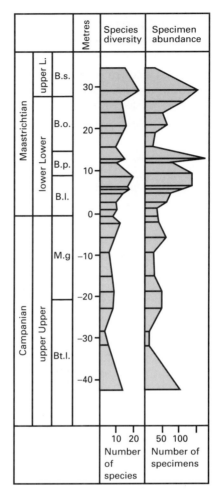

Fig. 5.2 Observed phenon diversity of late Cretaceous brachiopods plotted against relative abundance of specimens. Data comes from a section at Hemmoor, north west Germany (Surlyk, 1982).

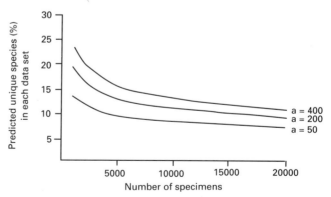

Fig. 5.3 The predicted percentage of unique species in each of two samples of equal size drawn from the same large population plotted as a function of sample size (Koch, 1987). a, Fisher's log series constant, which is proportional to data set diversity.

diversity of a rock unit and the number of workers publishing on that rock unit (Sheehan, 1977), indicating yet another factor that can generate sampling bias.

The best method for redressing biases in phenon abundance, when comparing two or more samples, is rarefaction (Tipper, 1979; Koch & Sohl, 1983). Rarefaction works by reducing the size of all samples to that of the smallest by predicting the number of species that would have been found at that sampling level. This requires collection data on the number of specimens recorded for each phenon in each sample.

Often, though, such detailed sampling information is simply not available. Koch (1987, 1991) summarized other ways in which sampling bias can be assessed where collection data are lacking. For example, the number of specimens collected in each of two areas allows an estimate of the percentage difference in taxonomic composition that is likely to be the result of sampling artefact alone.

Before any observed difference in taxic abundance between regional or stratigraphic samples is considered worth investigating, the null hypothesis – that the difference has arisen simply through sampling bias – must be rejected. Carter & McKinney (1992) used this approach in their comparative study of taxic diversity patterns of Eocene echinoids between two regions in south-eastern USA. They observed that, of the 40 phena known, a total of 35 are recorded from the northern region and 31 from the southern. Furthermore, sampling was not uniform, with approximately twice as many specimens known from the southern region. They therefore used Koch's (1987) technique to calculate the expected percentage difference in taxa attributable to unequal sampling between the two regions. This showed that about five taxa (12% of the total) might be expected to be unique to one area because of sampling error alone. Given that the observed difference was nine taxa, Carter & McKinney could reject the null hypothesis and examine other possible causes for the observed taxonomic difference.

Geographical range. A phenomenon also related to sampling is that:
> taxa having long geographic ranges are more likely to be preserved in the fossil record, because they occur at a greater number of potential fossilization sites. Therefore species with broad geographic ranges may have a higher probability of displaying longer geologic durations than taxa with narrow geographic ranges, even when no difference exists (Russell & Lindberg, 1988, p. 323).

Pease (1988d) adopted a mathematical approach to biases in the range of a taxon's duration, whereas Russell & Lindberg (1988) used computer simulations. Both approaches demonstrated that species with restricted geographic ranges will display shorter geologic duration *even when no difference in duration exists* (Fig. 5.4). High levels

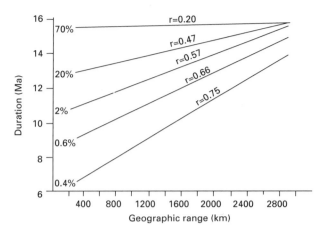

Fig. 5.4 The predicted duration of a taxon as observed in the fossil record plotted as a function of its geographic range, based on computer simulation (Russell & Lindberg, 1988). Each line is a regression line derived from assuming different levels of sampling completeness, from 0.4% to 70%. All taxa have the same duration (16 Ma). r, correlation coefficients for the regression lines calculated.

of sampling reduce this problem but, even assuming 70% sampling completeness, there was a significant underestimate of the geologic duration of the rarer taxa.

Stratigraphic and geographic ranges of late Cretaceous gastropods from south-eastern USA have been shown to be correlated (Jablonski, 1986a), but is this for reasons other than sampling? Marshall (1991) has given a clear and detailed treatment of this question and found that the correlation between geographic range and phenon duration appeared to be real and that there was a genuine difference between planktotrophic and non-planktotrophic phenon durations. However, his calculations are strongly dependent upon his measure of fossilization potential/km (f), which he calculated as total number of localities/Ma/km sampled. This simply measures the density of sampling used, and takes no account of possible biases in relative abundance. If geographically restricted phena also tend to be locally rare, then there will be a systematic bias to f and the observed correlation could still be an artefact.

It is important to recognize that sampling alone can give rise to an autocorrelation between species abundance, geological range, and geographic range.

Biomineralization and preservation potential. Between 50% and 95% of species in present-day marine communities have no skeletal component and are thus only very rarely preserved (Parsons & Brett, 1991). Of the remainder, the composition and structure of their skeletal elements directly affects their preservation potential. Taxa with fragile or spicular mineralized skeletons, or non-mineralized cuticles, obviously have very poor preservation potential. Even among relatively well-mineralized organisms, there are obvious differences in preservation potential. The mineralogy of skeletons strongly influences how they are affected during diagenesis (Dullo & Bandel, 1988; Carter, 1990; Tucker, 1991). Differential resistance of shells to dissolution is controlled by shell structure and composition, and can lead to a taxonomic bias in preservation potential (Peterson, 1976; Flessa & Brown, 1983; Koch & Sohl, 1983; Davies *et al.*, 1989). It is well known that aragonitic shells are more prone to dissolution than calcitic shells (Dullo & Bandel, 1988). For example, ammonite aptychi (which are calcitic) may occur in relative abundance in certain facies, whereas their phragmocones (originally aragonitic) are entirely wanting or preserved only as ghosts (Arkell, 1957).

Shell microstructure and composition can also significantly affect the resistance of shells to fragmentation and abrasion (Parsons & Brett, 1991). Thick-shelled oysters, with their robust and dissolution-resistant calcitic valves, stand a better chance of being preserved in the fossil record than, for example, the thin, dissolution-prone aragonitic shells of tellinacean bivalves (Jablonski, 1988). In arthropods it is the degree of calcification of the exoskeleton that controls preservation potential (Speyer, 1991).

Fragmentation of multiskeletal organisms is also a major problem, since disarticulated material cannot be identified to the same degree of taxonomic resolution as fully articulated specimens. This problem is particularly acute in echinoderms and, to a lesser extent, arthropods. Asteroids have a very poor fossil record in comparison to echinoids because their skeleton is membrane-embedded and falls apart at death, whereas the test of an echinoid is tessellate and holds together after death, sometimes for very long periods (Donovan, 1991). In a study of echinoid preservation Kidwell & Baumiller (1990) found that taxonomic differences in test rigidity (generated by degree of sutural interlocking between plates) correlated with how well tests were preserved.

In the plant record there are also likely to be strong preservational biases related to their composition, as shown by Tegelaar *et al.* (1991). They demonstrated experimentally that higher plant phena with a considerable amount of cutan, as opposed to cutin, in their cuticular matrix stand a much better chance of being preserved in the fossil record.

Taphonomic control on exceptional preservation faunas (Konservat-Lägerstatten) represents an extreme example, and Briggs & Gall (1990) found that taphonomic factors were so important in controlling the range of taxa preserved in such deposits that stratigraphic age had little influence in determining their taxonomic similarity.

Habitat. The environment in which a species lives directly affects its chances of preservation. This is largely because the chances of an individual becoming buried is largely dependent on sedimentation regimes, and these vary widely between environments. It is generally acknowledged that the highest chances of preservation are in marine sediments, but, even here, there are great differences in preservation potential between different habitats. For example, the marine fossil record of invertebrates is almost exclusively composed of continental shelf faunas. True deep-sea fossil faunas are extremely rare; one example is the Japanese Miocene deep-sea fauna of the Morozaki Group (Mizuno, 1991).

There is likely to be a bias against species dwelling in open, shallow-water, rocky habitats, since these are areas of net erosion where shells are likely to be broken up and pulverized (Parsons & Brett, 1991). Even in nearshore sedimentary environments, storm erosion and re-deposition predominate, making it likely that only the more robust skeletons will survive into the fossil record. Preservation is to a large extent determined by the immediacy, thickness, and permanence of burial. Anoxic conditions may (Allison, 1988) or may not (Plotnick, 1986; Kidwell & Baumiller, 1990) decrease the rate of decay, but play a primary role in promoting early mineralization, and in preventing scavengers from dismembering carcases on the sea-floor.

There may also be a latitudinal or palaeobathymetric gradient in the preservational potential of multiskeletal organisms linked to temperature. Kidwell & Baumiller (1990) reviewed the preservational potential expectation of echinoids in relation to both water depth and latitudinal temperature (Fig. 5.5). They argued that the best preservation would be expected in the zone lying below fair weather wave-base and above the wave-base of average storms (i.e. approximately 20–50 m water depth).

			Latitude		
			Low	Mid	High
Approximate water depth (m)	0 – 20	Protected coast and 'lagoons': rare but thick storm deposition ± anoxia	Poor	Poor to low	Moderate to good
		Open coast and nearshore: continuous fairweather and storm reworking	Worst	Poor	Poor to low
	20 – 50	Transitional zone: episodic storm deposition predominates	Low to moderate	Moderate to good	Best
	50 – 200	Open shelf: rare, thin storm deposition, ± anoxia	Low to moderate	Low to moderate	Moderate to good

Fig. 5.5 Predicted preservation potential for echinoids as a function of latitude and water depth (Kidwell & Baumiller, 1990).

Factors extrinsic to taxa

Lithofacies variation. The stratigraphic succession at many levels displays a mosaic of changing facies representing different habitats. Since all species distributions are to some extent environmentally restricted, some much more than others, the distribution of phena in the fossil record is typically very strongly correlated with lithofacies distribution. Thus, taxa often appear and disappear from the fossil record simply because of local immigration and extinction as facies distribution changes. Taxonomic range terminations can erroneously appear to be abrupt, simultaneous truncations because of sedimentary incompleteness or facies shifts (Sadler, 1981; Dingus, 1984). Even in the apparently stable conditions of the deep sea, sediment condensation or hiatuses can truncate ranges (MacLeod & Keller, 1991). Conversely, taxonomic range terminations can appear as uncorrelated events spread over the stratigraphic succession when in fact they represent an abrupt extinction event smeared through incomplete sampling (Signor & Lipps, 1982). Because the relative abundance of phena varies, rarer taxa will be less well sampled and will consequently appear to have less complete ranges than the more common phena so that observed last occurrences of rarer taxa are spread backwards in time from a single extinction event.

The importance of lithofacies bias has recently been highlighted by Carter & McKinney (1992). They used cluster analysis to discriminate two regionally distinct faunas of Eocene echinoids in south-eastern USA. The dissimilarity in taxonomic composition between the two regions was greater than expected from sampling biases alone (see above). However, they found that the differences were largely the result of temporal and lithofacies mismatch between the two areas. Terrigenous sand facies dominated in the northern region, and carbonate sands in the southern region. When the faunas of the two regions were averaged over a long geological timescale, significant differences in composition were apparent. However, a more careful comparison of samples from similar lithofacies and time zones suggested that: (i) much of an entire biozone represented in Florida was missing from the northern region; and (ii) where similar facies were developed in both regions during the same interval, the faunal similarity was very much greater. Carter & McKinney thus concluded that both stratigraphic and lithofacies mismatch were generating much of the observed faunal difference between the two regions.

Post-depositional processes. After fossilization, geological processes can only act to degrade the quality of the record. Reworking of older deposits can lead to a mixing of faunas of different ages. Erosion removes fossiliferous deposits from the record and acts at many scales from major regional events down to the microstratigraphical level

(Sadler, 1981). Several workers have attempted to quantify the completeness of the sedimentary record through analysis of gaps and hiatuses (Schindel, 1980, 1982; Sadler, 1981; Anders *et al.*, 1987; Allmon, 1989; Sadler & Strauss, 1990; McKinney, 1991). This work has emphasized that the sedimentary record often contains more gap than record. Sedimentary condensation and hiatuses are often highly significant and cannot be ignored (MacLead, 1991). Another important process is diagenesis, which tends to destroy fossils, often selectively removing all but the most robust and chemically stable skeletonized taxa. Regional tectonism is a large-scale expression of this process that wipes out the fossil record.

All these factors result in a general reduction in the amount of sedimentary rock available for study. This in turn leads to other biases. The area of surface outcrop of sedimentary rock and the volume of sedimentary rock are both strongly correlated with phenon diversity and increase towards the present (Raup, 1976a,b). There is a similar correlation between non-marine outcrop area and phenon richness of fossil plants (Tiffney, 1981). Although some of this may be no more than a function of small sample size (Signor, 1978, 1982), outcrop area remains a potential source of bias.

The importance of geological processes in the degradation of the fossil record has been highlighted in a comparative study of the Recent and Pleistocene marine molluscs of the Californian region (Valentine, 1989). Valentine estimated that the Pleistocene fossil record for the region was based on about 20 000 individual records of phena. Taking into account biases due to sample size differences, this suggests that about 85% of phena recognized today are preserved in the fossil record. Furthermore, the bulk of the Recent phena that are missing are those that are either deep-water or are small and fragile. Thus the percentage of shallow-water phena that become fossilized is probably even higher. This suggests that the fossil record is not composed predominantly of special event accumulations of shells, but rather that it represents the remains of an excellent original record formed mainly by standard episodic events. Valentine concluded that most durably skeletalized taxa enter the fossil record initially, but that this record is heavily depleted through the geological processes of erosion, burial, and *in situ* destruction of skeletal elements through time.

Palaeogeographic distribution of well-studied areas. One final bias arises from the distribution of well-studied areas in palaeogeographic space. There is undoubtedly a very strong sampling bias towards western Europe, North America, and China, where there has been a long tradition of taxonomic palaeontology and biostratigraphy. In Raup's (1976a) phenon diversity data, 54% of the phena come from North America and Europe, yet this represents only some 16% of present-day land-surface area. There is an even greater bias in groups such as Triassic–early Jurassic echinoids, where more than 90% of phena come from these two areas (Smith, 1989). Allison & Briggs (1993) have pointed out that shifting palaeogeographic configurations may have biased estimates of phenon diversity through time. Phenon diversity shows a latitudinal gradient, being greater at low latitudes than at high latitudes, therefore temporal bias in global diversity estimates could have arisen from changes in the latitudinal placement of continents through time. Allison & Briggs point out that almost all marine sedimentary rock in the Palaeozoic of Europe and North America was deposited in tropical latitudes, whereas only 24% was deposited in these latitudes during the Mesozoic and Tertiary. Thus a high percentage of the palaeontological database represents a relatively limited geographic area which, through time, has shifted in latitude. This will surely have affected our perception of global patterns of evolution.

Estimating absolute taxonomic ranges

Sampling and preservation are significant problems in reconstructing trees from biostratigraphic data and observed ranges are likely to underestimate true ranges. The goal for the biostratigrapher is to decipher the true ranges of taxa from patterns of local appearance and disappearance generated by shifting facies and sampling artefacts. Two approaches have been developed that improve biostratigraphic inference. These encompass, on the one hand, quantitative biostratigraphic correlation methods such as graphic correlation, and on the other, methods of placing confidence intervals on observed taxonomic ranges. The former concentrate on constructing the most complete set of taxonomic ranges by combining data from different sections; the latter focus on estimating single taxon ranges from sampling distribution data. A third method, which uses phylogenetic analysis to constrain taxonomic ranges, is discussed in Chapter 6.

Quantitative biostratigraphic correlation

This field has developed enormously in recent years and is covered in detail by Cubitt & Reyment (1982) and Gradstein *et al.* (1985); excellent summary reviews are given by Edwards (1989, 1991). There are various quantitative methods that integrate data from two or more sections into a composite reference section providing the best estimate for the actual sequence of taxonomic appearances and disappearances. It is important to use a method for range zone estimates and chronostratigraphic estimation that discovers the maximum range of a taxon. Thus, Hay's (1972) probabilistic method, which provides high-level correlation but tends to underestimate taxonomic ranges, is inappropriate. Graphic correlation, which was initially developed to make biostratigraphic correlation more rigorous (Shaw, 1964), offers a more appropriate approach.

The method starts with a measured section on which the positions of first and last appearances of taxa are recorded. A second measured section is then compared to the first by placing each section along an axis of a graph (Fig. 5.6). First appearances (FA) and last appearances (LA) of all taxa common to both sections are then plotted but kept discrete. Once this has been done, the best supported correlation is obtained by drawing a line through as many FAs and LAs as possible. Since FAs can only be equal to or later than the true first appearance, and LA can only be equal to or earlier than the true last appearance, the correlation line should be positioned so that LAs preferably lie on or below the line and FAs lie on or above the line. Note that the

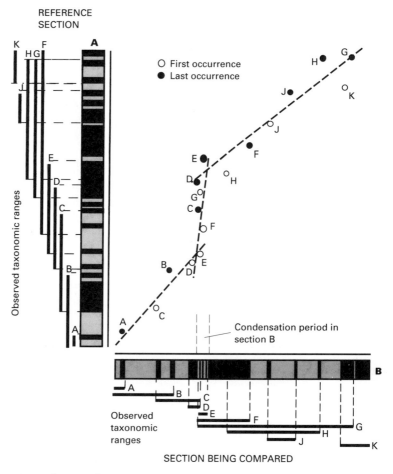

Fig. 5.6 Graphic correlation – hypothetical example to show the way in which an apparently abrupt faunal turnover in section B can be recognized to be the product of sedimentary condensation by graphic correlation against a standard section. The ranges of taxa A–K are indicated for each section. Dashed lines indicate the best-fit correlation between sections.

correlation need not be a straight line, since a change in slope indicates a change in relative sedimentation rate.

Having determined the best line of fit, anomalous LAs and FAs can be identified and, by projecting these onto the correlation line, their estimated true position can be identified on the section. This section then becomes the standard reference section for comparison with other stratigraphic sections. Since the order of comparison of the sections might affect the ultimate outcome, graphic correlation should be carried out several times using different initial sections. In this way a number of potentially very incomplete sections can be amalgamated to produce an estimate of the complete sequence. This sequence is the one supported by the greatest number of taxonomic datum points. As in cladistic analysis of morphology, parsimony is the ultimate arbiter. Anomalous stratigraphic appearances or disappearances of taxa are identified on the strength of total evidence.

If there is significant scatter in the distribution of FAs and LAs then the observed stratigraphic ranges are likely to be incomplete in the sections studied. In some cases, however, the correspondence of FAs and LAs between sections is extremely good. Cooper & Lindholm (1990) carried out a global graphic correlation for 132 abundant and widespread phena of early Ordovician graptolites and obtained a remarkably good fit, implying that the inferred stratigraphic ranges approximate closely to the true total ranges for these phena.

More recently, graphic correlation has been expanded so as to utilize all potentially informative datum points (Dowsett, 1989; Edwards, 1989; MacLeod, 1991). Thus, by including magnetic polarity reversal horizons, ash bands, microspherule layers, stable isotope spikes, and other such chronostratigraphically useful markers in the graphic correlation, the resultant reference section can be made even more reliable.

Confidence intervals on taxon ranges

Paul (1982) first used gap-length frequencies between samples in order to assess the probability of a taxon's range extending beyond its observed first or last occurrence in a section. Later Strauss & Sadler (1987, 1989) formalized this method into a rigorous statistical procedure for placing confidence intervals on observed stratigraphic ranges, and this procedure has been used by Springer & Lilje (1988), Allmon (1989), Springer (1990), Marshall (1990, 1991), and McKinney (1991).

In its initial formulation Strauss & Sadler (1989) were primarily interested in calculating confidence intervals for observed taxonomic ranges in local (i.e. single) sections. Their method requires only the total observed range and the number of occurrences (fossiliferous levels) within that range. The confidence interval to be added to the end of a range is then calculated using the equation:

$$\alpha = (1 - C_1)^{-1/(H-1)} - 1$$

where α is the confidence interval expressed as a fraction of the total observed stratigraphic range, C_1 is the required confidence interval (e.g. 0.95, 0.99), and H is the number of observed fossiliferous levels (see Marshall, 1990, 1991). Confidence intervals can be quite large for small samples, and only approach around 10% of the observed range when 30 or more fossil horizons have been identified.

The method is not without its problems, because it makes a number of key assumptions that must be met if the results are to be valid: (i) it assumes random sampling – if fossiliferous horizons are not randomly distributed in the sequence (due to variation in facies, sedimentation rates, or collection intensity) then the calculated confidence intervals will be invalid; and furthermore, (ii) it assumes that identical facies and preservation potential continue at either end of the observed range, i.e. the appearance and disappearance of a taxon is not linked with a change in facies – as some facies changes can be very subtle, this is very difficult to test, but one way is to use the presence of similar taxa as negative controls.

In discussing the confidence intervals on phenon ranges of the Neogene bryozoan *Metrarabdotos* in the Caribbean, Marshall (1990) used the presence of other bryozoans of the same genus as a guide to whether suitable conditions were present at either end of the range for the preservation of the phena under study (Fig. 5.7). Confidence intervals are thus most appropriate in sections of monotonous lithofacies, such as graptolitic shales or chalks. Even here, though, care must be taken if the confidence intervals are given in thickness of sediment. Any significant fluctuation of sedimentation rate within the range of a phenon will generate a non-random distribution, and if there are significant hiatuses within the error bar interval, this interval will be overestimated.

The calculation of confidence intervals for observed stratigraphic ranges in single sections is relatively straightforward, even if some care is needed. The technique may also be generalized to taxonomic ranges in composite sections. However, although there is no ambiguity in correlating the first or last appearance of a taxon on a composite section, correlating individual sample levels within the range is much more problematic. If composite sections are to be used, they must be drawn up independently of the phena whose confidence intervals are to be calculated, since graphic correlation makes the position of one fossil horizon causally dependent on that of the others (Marshall, 1990).

Confidence intervals might also be applied to taxa higher than phena by using composite sections. This could be achieved, for example, by equating sampling with the presence/absence of a taxon in the zones encompassed by its total range. However, the assumptions required here are even less likely to be met: (i) it must be

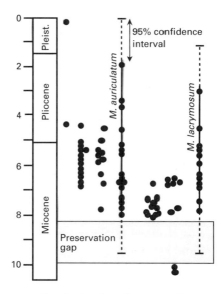

Fig. 5.7 Confidence intervals calculated on the range of two species of the Neogene bryozoan *Metrarabdotos* (Marshall, 1990, with data from Cheetham, 1986, 1987; see Fig. 2.2). The occurrences of other *Metrarabdotos* species are indicated and act as negative controls. Each circle indicates a species record at that stratigraphic level. Note that there is a period between about 10 and 8 Ma where no *Metrarabdotos* are found.

demonstrated that suitable facies occur in each of the zones, so that the absence of the taxon is not simply due to the absence of appropriate sedimentary environments. This could be done using other taxa of very similar ecology and preservation potential as negative controls; (ii) the zones are assumed to be approximately of equal duration; and (iii) the taxonomy is assumed to be consistent though not necessarily cladistic: confidence intervals can be calculated for paraphyletic and polyphyletic groups provided they are consistently recognized.

Finally, it must be stressed that confidence intervals should not be taken to indicate a taxon's total range. The method identifies the stratigraphic interval beyond which we can be statistically confident that a taxon's absence is not simply the result of *sampling* failure in appropriate facies; the method does not identify the interval for which we have statistical confidence that a taxon was present. It is therefore not a method for determining true range: the sudden appearance or disappearance of a taxon in a section may be a genuine event and not a sampling artefact, even though confidence intervals can be calculated beyond those points. Furthermore, estimated confidence intervals relate only to local extinction or immigration events if the studied sections do not include the true last appearance or first appearance of a taxon.

Taxonomic ranges: do they provide a test of phylogenetic hypotheses?

Both graphic correlation and confidence interval estimations are rigorous methods of quantifying the range of taxa based on the observed stratigraphic data. But if observed stratigraphic data are incomplete (for example, because of incomplete regional coverage) then these methods simply fail to identify total range. If, for example, a taxon originates and evolves outside the study area and only later migrates into it, then graphic correlation and confidence intervals will only relate to the latter part of that taxon's history. Confidence intervals at either end of the observed stratigraphic range could be quite small if sampling is relatively dense in the studied area, yet still fall grossly short of the total range of the taxon. The same applies to local extinction events and peripheral (unsampled) populations that survive beyond this. These methods therefore provide us only with an estimate of the maximum *observed* range for a taxon. They make no prediction about the total range of a taxon unless geographic sampling is extensive, which it rarely is. A better method of estimating total range comes from the construction of phylogenetic trees (see p. 139).

Paul (1982) developed an argument concerning the probability that two taxa with overlapping ranges might be preserved in the wrong order. Given equal probabilities of preservation and discovery, the chance of finding two taxa preserved in the wrong order, even when just one example is drawn from each, is never more than 50%. As sampling improves, so also do the chances that taxa will be observed in their correct order. Even when the probability of finding a specimen in the two taxa is inversely related through time. Paul argued that the worst possible case one might expect is for 25% of the ranges to be inverted. Given that most phena do not have overlapping ranges, Paul concluded that phena and taxa as documented in the fossil record are, in general, preserved in the correct order. This must of course be true, simply because the number of pair-wise comparisons of taxa with non-overlapping ranges greatly exceeds those with overlapping ranges in the fossil record.

Paul's argument is based upon the sampling error associated with observed ranges. With reasonable sampling there is little doubt that the actual order of taxa preserved in the section or sections will be correctly assessed in the vast majority of cases. However, if the sections available for study do not encompass the total range of the taxa, the observed distribution may still preserve taxa out of order. The question that needs addressing therefore, is not 'How good is our sampling of observed sections?' but 'Is our sample of geological sections good enough for documenting the order of taxa correctly?' The high consistency of results obtained from graphic correlation over wide geographic areas shows that sampling for at least some of the more common, stratigraphically useful, phena is adequate. But such stringent testing remains very much the exception rather than the rule.

Fortey & Jefferies (1982) used a different approach by examining how sampling density affected the likelihood that the order of appearances in the fossil record was correct. They constructed a model evolutionary tree and sampled taxa from it at random many times. On each occasion, they reconstructed an evolutionary tree using only those taxa sampled, and observed how often taxa appeared out of order in that tree.

From their findings they argued that if only a small percentage of all the phena that existed were sampled from a taxon, then there was a high probability that phena would be preserved out of order. In more heavily sampled taxa, however, the order of appearances in the fossil record would become much more concordant with that predicted from the phylogenetic tree.

Recently the correspondence between predicted order derived from phylogenetic analysis and observed order derived from the fossil record has been tested empirically by Norell & Novacek (1992a,b). Their initial analysis (1992a) used 20 studies of selected vertebrate groups and compared the order of appearance of taxa deduced from phylogenetic analysis with the observed order of appearance of the same taxa in the fossil record. They did this by plotting cladistic rank against stratigraphic rank for each dichotomy. They obtained very mixed results (Fig. 5.8). some groups, such as equids, show a very strong correlation and clearly represent a densely sampled taxon both in terms of phena and total geographic range. Other groups (e.g. higher primates), where sampling is clearly far from satisfactory or phylogenetic relationships are poorly constrained, show a very poor correlation. Overall 18 of the 24 cases examined showed a strong correlation between stratigraphic age rank and cladistic rank, implying that our sample of the fossil record, though far from perfect, does tend to preserve taxa in the correct order more often than not. For marine invertebrates with a

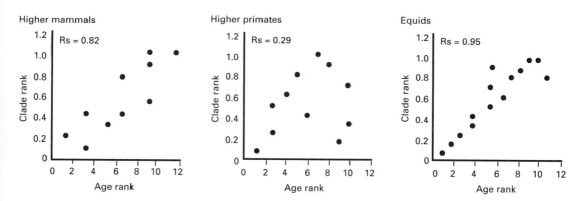

Fig. 5.8 Correlation of age rank with cladistic rank for three groups of higher vertebrates (Norell & Novacek, 1992a). Cladistic rank is determined from the number of nodes between the termini and the basal node. Age rank is determined from the first appearance of the taxon in the fossil record. Rs, Spearman rank correlation coefficient.

better fossil record the correspondence is likely to be even better. Where there is mismatch, either the phylogenetic analysis or the stratigraphic record or both are in error, and further tests are needed to identify the cause of the discrepancy.

Summary

The observed distribution of fossils in the geological record is our only direct evidence for the time of origination and extinction of clades. However, the fossil record is very imperfect and there are many biases that might lead to underestimation of taxonomic ranges. Biases act at all stages, controlling whether organisms enter the fossil record, whether fossils survive in the geological record, and whether they are discovered and recorded. Inadequate sampling may lead to significant underestimation of stratigraphic ranges of rare taxa or geographically restricted taxa. Taxa differ in their preservational potential because of differences in the degree of mineralization and the resistance of any skeletal hard parts to solution and abrasion. Preservation potential also depends upon the environment inhabited. Geological processes lead to ever more destruction of the fossil record such that the perceived diversity of taxa through time may be biased by the availability of rock outcrop of the appropriate age and facies, and by the palaeographic distribution of well studied land-masses in the geological past. For all these reasons, observed stratigraphic ranges of taxa may be significantly underestimated.

Graphic correlation and the placement of confidence intervals on the ends of ranges are techniques based on sampling theory that have been developed to estimate total stratigraphic range from distributional data. Nevertheless, sampling theory and graphic correlation only help to refine our estimates of taxonomic ranges based on observed distribution within available sections. Since available sections represent such a small percentage of the original area of deposition, and since most fossil taxa are rare anyway, it is by no means certain that the ranges that we can observe give an adequate representation of the true ranges of taxa. Therefore we need a technique that can predict taxonomic ranges and is not tied to sampling theory based on observed distributions. Such a method is provided by the construction of phylogenetic trees (Chapter 6).

6 The construction of evolutionary trees

The construction of evolutionary trees brings together two independent lines of evidence: systematics and biostratigraphy. Chapters 2–4 examined the problems involved in recognizing and grouping phena, and how their phylogenetic relationships can be established through cladistic analysis of morphological data. Chapter 5 introduced the other crucial source of data, the biostratigraphic record. The present chapter addresses how phylogenetic relationships and biostratigraphic evidence can be combined in the construction of trees.

Trees, cladograms, and ancestors

The fundamental distinction between cladograms and trees is that cladograms are concerned only with the establishment of discrete nested relationships of taxa through formal character analysis, whereas trees are concerned with establishing serial continuity among taxa through hypotheses of ancestry and descent. The cladogram specifies atemporal sister group relationships on the basis of homologies shared at the correct level of inclusiveness (synapomorphies). It identifies a hierarchy of taxa within taxa, each of which can be unequivocally recognized on the basis of some uniquely shared homology. A tree specifies hypotheses of ancestor–descendant relationships by adding the dimension of time. Trees contain more inference than cladograms, but are essential for improving our understanding of evolutionary patterns, as will be demonstrated in the next chapter.

There are two conventions for transforming a cladogram into a tree (Eldredge, 1979; Patterson, 1983). Firstly the cladogram can simply be calibrated against the fossil record. The geological ranges of terminal taxa are plotted against the stratigraphic column, and sister groups are linked so that they arise simultaneously either at or immediately prior to the first appearance of the stratigraphically older sister group. In this way no terminal taxon in a cladogram is identified as a direct ancestor to any other and the concept of ancestry remains purely notional. Such trees have been termed X-trees.

The alternative is to construct A-trees, in which some of the taxa in the cladogram are placed as direct ancestors to other taxa. A-trees imply that actual ancestor–descendant relationships can be discovered, or at least formally hypothesized.

Which approach is adopted clearly depends on whether ancestors are considered to be discoverable. It is therefore necessary to explore the concept of ancestry in more detail.

The concept of ancestry

Ancestry can be conceptualized in one of two ways: taxon A may be said to be the ancestor to taxon B, or taxon A may be said to include the ancestor of taxon B. Both run into problems when applied to the fossil record.

To say that one taxon gives rise to another is simply a syntactical error. No taxon gives rise to another taxon other than in the sense that differentiation occurs through time and a taxon gives rise to subtaxa. Genera do not give rise to other genera. Even at species level, Cracraft (1989, p. 47) and Nelson (1989b) have argued that species do not give rise to other species; it is differentiation among populations that gives rise to species. Widespread species become dispersed and partitioned through isolation (either physical or behavioural) and it is isolated populations that give rise to new species. Species arise from among the units at the next level down in the hierarchy. Thus species are speciated, they do not themselves speciate. Furthermore, higher taxa arise coincidently with their basal subtaxa. The origin of a family coincides precisely with the origin of its first included genus, which in turn coincides with the origin of its first included species taxon.

What of the more general statement that one taxon originates from within another, older taxon? Here the older taxon is not the ancestor *per se* but potentially includes subunits which can be considered ancestral. The problem with this concept stems from the validity of the ancestral taxon and has been dubbed the 'disappearing basal syndrome' by Nelson (1989b). An ancestral taxon cannot be recognized by any unique characters that it has, only by the fact that it *lacks* the unique derived characters of its presumed descendant. Any taxon defined in this way must be paraphyletic and thus artificial both in theory and practice. Basal ancestral taxa are thus extinct paraphyletic groups and inevitably disappear from classifications as they become better studied and understood. For example, thecodonts were for a long time thought to be ancestral to an array of primitive archosaurian reptiles, but recent cladistic analyses have shown that thecodonts are a paraphyletic group, and thus cannot be considered to have any objective reality.

Can the notion of objective ancestral groups be saved by reducing the problem to species level? Throughout this book I have argued that the basal plesiomorphic taxa recognized in the fossil record (metaspecies) are grades which have arisen by default because of insufficiency of evidence. Metaspecies in the fossil record do not therefore represent hierarchically indivisible basal units of evolution. Thus, even fossil metaspecies, if they are ancestral, must be paraphyletic and established by convention.

So the concept of ancestry remains problematic. Taxa do not give rise to other taxa; they arise by differentiation from taxa at the next lowest level in the hierarchy. Any group hypothesized as ancestral to another is recognized on the basis of negative data (the absence of homologies that define the presumed descendant). Such groups must be paraphyletic if they give rise to other taxa, and are thus artificial constructs.

A pragmatic approach to ancestry

Although the identification of ancestors as objective taxa cannot be justified, potential 'ancestral groups' are in practice an inevitable constituent of all cladistic analyses. Even when all demonstrably polyphyletic and paraphyletic taxa are removed from a cladistic analysis, the terminal taxa will comprise a mixture of monophyletic taxa and operationally indivisible plesiomorphic taxa (i.e. plesiomorphic taxa that cannot on available evidence be subdivided). These metataxa are grades, and may be paraphyletic. As such they could include ancestral populations to other derived taxa.

Many so-called fossil species (phena) are surely metataxa, plesiomorphic groupings that may or may not be resolved more finely given additional morphological information. Others are monophyletic. The same combination of metataxa and monophyletic taxa are likely to be encountered at any level in the hierarchy during the initial stages of a cladistic analysis of a group. As knowledge of a group increases and more detailed character analyses are performed, metataxa will be broken into smaller units. These, however, will still consist of both minimally diagnosable grades and monophyletic taxa. Phenon-level trees differ in no material way from trees constructed from higher taxa.

What is to be done with metataxa when transforming a cladogram into an evolutionary tree? There appear to be two options: (i) we can construct X-trees based on the distributions of demonstrably monophyletic taxa, ignoring metataxa; or (ii) we can construct A-trees using all available taxa and placing plesiomorphic basal taxa as ancestral groups where stratigraphic evidence suggests this is possible. Ignoring metataxa may result in a large proportion of taxa being omitted from any tree reconstruction. This in effect throws away useful information about the time of origination of any higher taxon that includes the basal plesiomorphic grade. Therefore the approach advocated here is to use both monophyletic taxa and operationally delimited metataxa in tree construction, but to use a convention to distinguish between the two, since only the former are likely to survive more detailed analysis.

What does it mean when a metataxon is placed in direct stratigraphic continuity with another taxon? The interpretation must be that the metataxon (be it a phenon or some higher-level taxon) potentially includes one or more populations that have given rise to the derived taxon. If that is the case, the metataxon is paraphyletic and further work is required to resolve relationships within it. Thus all metataxa placed in direct lines of descent represent artificial groups that are necessities imposed by limitations of available evidence. They are not ancestors *per se*, even though their membership may include ancestral populations.

How phylogenetic trees are constructed

The construction of a tree begins with the study of the distribution of morphological characters within the taxa. From this a character–taxon matrix can be assembled and a cladistic analysis carried out. The resulting cladogram provides a hypothesis of relationships that is independent of biostratigraphic information. If this is tested and branches are found to be well supported, it is worth proceeding to convert the cladogram into an evolutionary tree. If the cladogram is only weakly supported then a tree can still be constructed, but there will be less certainty that any mismatch between stratigraphic and phylogenetic order is due to failings of the fossil record.

The branching pattern derived from the analysis of morphological data can now be used to connect the observed biostratigraphic ranges of sister group taxa in a way that requires the minimal amount of hypothesized range extension. Range extensions are *ad hoc* assumptions that have to be made about gaps in the fossil record in order to make the biostratigraphic and phylogenetic evidence concordant. In effect they are periods of geological time for which we have to assume a taxon existed even though there is, as yet, no direct evidence for it in the fossil record. Since range extensions are purely conjectural, they should be kept to a minimum, otherwise any possible branch length extension could be invoked at whim.

Taxa that are demonstrably monophyletic (i.e. have their own set of synapomorphies) are assumed not to be ancestral to any group without those synapomorphies. Otherwise, if reversal is permitted, then anything in principle can be designated ancestral to anything else and the floodgates are opened for accepting any scenario. Ancestry is permissible only for plesiomorphic basal taxa.

In constructing trees, therefore, three assumptions must be made:
1 The morphological cladogram is well supported and provides the best current estimate of phylogenetic relationships.
2 Demonstrably monophyletic taxa have not given rise to other taxa.
3 Stratigraphic range extensions should be kept to a minimum.

It is simplest to start by considering cases where all terminal taxa are phena (the problems and pitfalls of using higher taxa as terminal units are discussed on p. 137). I shall also assume that all the known members of a taxon are well documented and have been included in the cladistic analysis as terminal units. The way in which range extensions are added to observed biostratigraphic distributions depends upon the nature of the terminal taxa, in particular whether they are plesiomorphic (and thus metataxa) or have their own apomorphies.

The following three-taxon examples encompass the range of possible permutations.

When all phena are demonstrably monophyletic (Fig. 6.1)

For three taxa a fully resolved cladogram with all branches supported by apomorphies requires a minimum of four characters. In the example shown, phena B and C are identified as sister groups to the exclusion of A. The observation that B has its own unique derived character debars it from being the direct ancestor to C, and the same argument can be used to debar C from being the ancestor to B. If these two taxa are sister groups, they must have split before either acquired their unique apomorphies. The smallest amount of range extension is added by assuming that the first appearance of either B or C marks the timing of this split and the acquisition of the apomorphies of both groups. If both B and C appear simultaneously in the fossil record then there is no need to extend the range of either. If, however, one of the

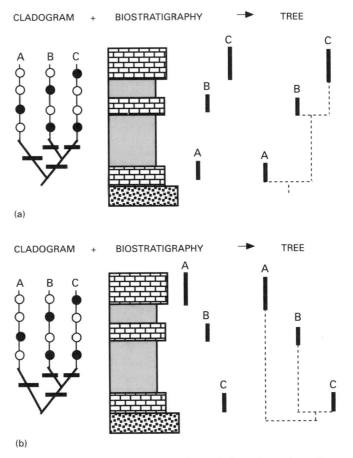

Fig. 6.1 Construction of phylogenetic trees from cladistic data where all taxa are supported by apomorphies (see text).

two appears later in the fossil record than the other (in Fig. 6.1a, C appears after B, whereas in Fig. 6.1b, B appears later than C) then the taxon appearing later in the fossil record needs to have its range extended back to the time of first appearance of the older taxon. This is indicated in the figures by the addition of a dashed line.

Exactly the same logic can be applied to unite the sister groups A and (B+C). Again both groups have their own unique apomorphies and thus cannot be considered as putative ancestors. The smallest amount of range extension results from assuming that the two taxa acquired their apomorphies at the time of their splitting and that this coincides with the first appearance of one or other in the fossil record. If A predates (B+C), as in Fig. 6.1a, then it has to be assumed that the common stem of taxon (B+C) extends from the first appearance of B back to the first appearance of A, since this is the very latest time that the two taxa can have diverged. Any other assumption will increase the amount of range extension that must be added. For example, the assumption that B and C diverge halfway between the appearance of A and the appearance of B will require range extensions of both B and C in place of a single stem lineage.

Notice that the tree topology (though not the branch lengths) remains identical to that of the cladogram, irrespective of the stratigraphic order of the taxa, when all taxa have their own apomorphies. Thus, in Figs 6.1a&b, taxon A is always sister group to taxon (B+C) and the divergences are set by the first appearance of one or other sister taxon in the fossil record.

When there are plesiomorphic phena (Fig. 6.2)

The situation becomes more interesting when some phena are metataxa and thus plesiomorphic with respect to their sister group. A resolved cladogram in which all three taxa are morphologically distinct can be achieved with only two apomorphic characters. One character defines (B+C) as derived compared to A and the other distinguishes C as derived from B. Thus phenon A is entirely plesiomorphic with respect to taxon (B+C) and phenon B is wholly plesiomorphic with respect to phenon C. Where the plesiomorphic phenon comprises individuals from a single sampled horizon it is probably correct to think of this as equivalent to a biological population. As such it can be regarded as potentially ancestral to the derived taxon, since all its derived characters are also present in its sister group. This does not of course prove it to be an ancestor, since ancestors cannot be recognized on any morphological criteria (see above). However, with the addition of biostratigraphic data and working on the principle of minimizing *ad hoc* range extensions, the fewest assumptions are made by postulating that the origin of the derived group lies within its plesiomorphic sister group.

Where the plesiomorphic phenon is composed of two or more discrete populations grouped together it represents a metaspecies, which may be monophyletic or paraphyletic but is beyond current limits of resolution. As we are dealing with samples that are dispersed in time and space, we cannot assume that our inability to resolve relationships is due to there being only tokogenetic relationships amongst the included individuals. Instead we must use other criteria to decide if the

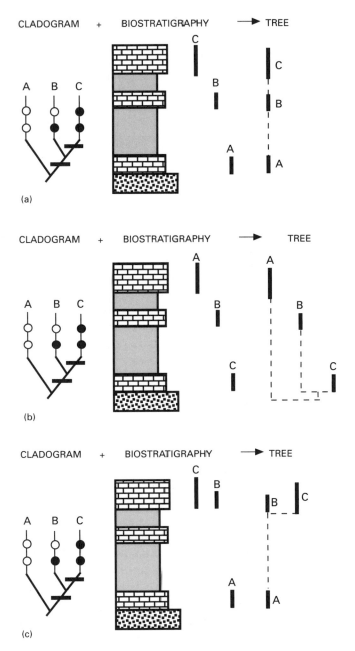

Fig. 6.2 Construction of phylogenetic trees from cladistic data where some taxa are plesiomorphic (see text). *Continued on p. 132.*

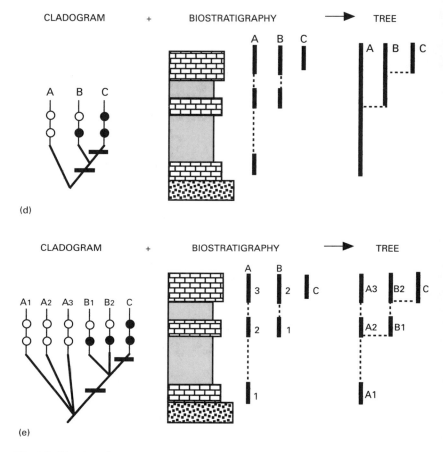

Fig. 6.2 *Continued.*

taxon is likely to be monophyletic or paraphyletic. Here stratigraphy can offer the necessary evidence.

In the first case (Fig. 6.2a) phena A, B, and C succeed one another with no overlap in the stratigraphic record. In this situation the minimal range extension is achieved by considering A, B, and C as part of a single evolving complex. Phenon A differentiates to give rise to phenon B by the acquisition of a derived character (the one that unites B and C), while phenon B further differentiates at a later stage to give rise to phenon C by the acquisition of the second apomorphy. Range extensions are therefore required only to fill in the gaps between the three taxa. The acquisition of the first derived character state therefore occurred at some point between the last appearance of phenon A and the first appearance of phenon B.

If, however, the stratigraphic distribution of the three phena is reversed such that phenon C is the oldest and phenon A the youngest then there are no grounds for considering any of the observed taxa as ancestral to any other. Phenon C has its own unique apomorphy and thus cannot either in part or in total give rise to its sister taxon B without assuming reversal. Since B appears in the fossil record later than C, it is necessary to extend the range of phenon B back to the origin of C at the very least (Fig. 6.2b). Similarly the range of phenon A needs to be extended back to the first appearance of taxon (B+C), since (B+C) cannot be direct ancestor to A. Thus here we can assume that both metaspecies are really monophyletic.

Where phenon A predates both B and C and these latter both appear simultaneously in the fossil record (Fig. 6.2c), the minimal range extension is achieved by considering A as a metaspecies from which both B and C potentially differentiated. No range extension is required to unite B and C as sister taxa, since they both appear simultaneously. As in the first case, phenon A can be transformed into phenon B simply by the acquisition of a single derived character state. The solution that requires the fewest assumptions then treats A and B as a single lineage from which phenon C branches. The apomorphy that unites B and C could have arisen at any time between the last appearance of A and the first appearance of B, and this sets the earliest possible time for the appearance of phenon C. However, in order to minimize the amount of range extension, phenon C is assumed to originate at its first observed occurrence in the fossil record.

In the final example (Fig. 6.2d) phena A and B are represented at more than one discrete stratigraphic level, and all three taxa have overlapping ranges. Both A and B represent metaspecies erected on the basis of plesiomorphy alone and are thus of unknown phylogenetic status. If A appears in the fossil record prior to B, and B appears prior to C, then direct ancestor–descendant relationships can still be postulated, but only by assuming that A and B are both paraphyletic concoctions of two or more biological populations. In this case no *ad hoc* range extensions are necessary. However, phenon A and phenon B must both represent groupings that have been created by error. A more accurate representation of the relationships amongst taxa A, B, and C is shown in Fig. 6.2e. Metataxa, which are simply groupings of polychotomies by taxonomic convention, are indicative of no more than our inability to resolve relationships at lower levels. In this case, biostratigraphic data have helped to resolve the nature of these metataxa by indicating that they are most probably paraphyletic.

Hypotheses of ancestry arise when morphological and biostratigraphic evidence are combined and the assumptions that have to be made about character change and gaps in the fossil record are minimized. Furthermore, groupings founded on plesiomorphy (metataxa), at the limits of available morphologic resolution, do not necessarily correspond to true biological units, but rather are groupings of taxonomic convention. It is therefore extremely important to identify whether terminal units on a cladogram are monophyletic or not, as this affects how tree branching is reconstructed.

Unresolved trichotomies (Fig. 6.3)

Sometimes three-taxon problems cannot be resolved and a trichotomy is formed in the cladogram. These trichotomies may be the result of conflicting evidence or simply reflect a lack of informative characters. Trichotomies that arise from conflicting data cannot be used in tree construction and must be resolved before proceeding (see below). However, trichotomies that result simply from a lack of informative characters can be used. Such a situation might arise where two of the three taxa each have a single apomorphic state while the third (A) is plesiomorphic. In this case taxon A may be a putative ancestor if it predates one or both of the other taxa.

If taxa A, B, and C are phena based on single populations that succeed one another without overlap (Fig. 6.3a) then neither B nor C can be ancestral to the others since they both have derived morphological attributes. The two phena must therefore be connected as shown, with the range of C extended backwards to the first appearance of B. Phenon A, however, has no derived characters unique to it and can thus be considered as a putative ancestor to both B and C. The fewest assumptions are made by treating A as ancestral to both B and C, and having B and C originate at the first appearance of B.

The situation in Fig. 6.3b is very similar except that phenon A is found at both the start and end of the succession. A is thus composed of two discrete populations, either one of which could be a possible ancestor to both B and C, while B is debarred from being ancestral to C because of its apomorphy. The simplest solution is to assume that A is a paraphyletic group and to derive B, and C from A[1] independently. In this way no range extensions need be evoked.

When all three phena in a trichotomy have their own unique derived characters (Fig. 6.3c) then none can be considered as potentially ancestral and tree construction proceeds by uniting the two more recent phena as sister group to the oldest of the three (in this case A). Note that the order of phena in the stratigraphic record makes no difference, since the cladogram provides no constraint on branching order. The stem branch that is hypothesized (here the branch leading from the split of B and C to the base of A) is assumed to be plesiomorphic with respect to all three taxa.

Pitfalls to be avoided

Beware of consensus trees. The construction of trees from trichotomies as outlined above is only valid if the three taxa involved are united because of absence of data. If the trichotomy is a result of conflict in the data set, the consensus tree is not necessarily compatible with any of the possible branching patterns it purports to summarize and none of the constituent cladograms may actually correspond to the consensus topology (Swofford, 1991). Consensus trees are a means of

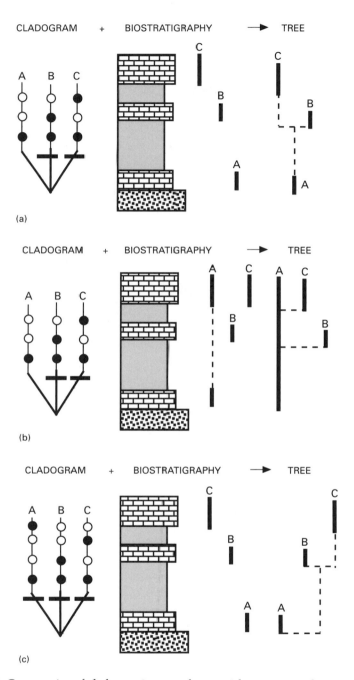

Fig. 6.3 Construction of phylogenetic trees where a trichotomy exists because of insufficient resolution.

showing areas of cladogram topology that are in agreement, while removing taxa that show conflicting positions to their lowest uncontested level in the hierarchy. A consensus tree is thus not a suitable basis for a classification or phylogenetic hypothesis (Miyamoto, 1985;

Carpenter, 1988) and should not be used to construct a phylogenetic tree. Instead, other data should be sought, or characters weighted to select from among the equally parsimonious solutions (Fig. 6.4). Stratigraphic order is one possible means for selecting from among equally parsimonious solutions.

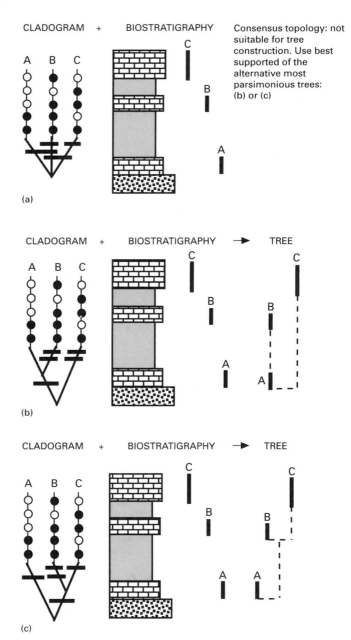

Fig. 6.4 Construction of a phylogenetic tree where there is character conflict in the data set.

Beware of higher taxa. The method outlined above can be applied equally well to higher-level taxa but much more care and attention is required if trees are to be accurate. The danger is that the characters attributed to a higher taxon are often based on a small number of the better known phena and may not hold for all phena assigned to that taxon. If less-well known or more incompletely preserved phena are included in the higher taxon, there is the distinct possibility that these may falsely extend the stratigraphic range of an apparently mono-phyletic taxon. For example, in the case of asaphine trilobites dis-cussed below (p. 144), the family Asaphidae is identified by Fortey & Chatterton (1988) as monophyletic because it has a derived character, the supramarginal course of the dorsal sutures in front of the glabella. This taxon has a long geological record but it is first recorded from the latest Middle Cambrian of Australia (*Griphasaphus griphus*). It is on the strength of this record that Asaphidae are extended back to the Middle Cambrian. However, *Griphasaphus griphus* is based on rather poorly preserved material and in fact the position of the dorsal suture in front of the glabella cannot be determined (R.A. Fortey, personal communication). Thus the extension of Asaphidae back to the Middle Cambrian is unjustified since its earliest member turns out not to show the derived character that distinguishes Asaphidae from its sister taxon.

It is therefore essential that the phena defining the earliest and latest records of any higher taxon are carefully checked to ensure that they show an appropriate suite of characters before accepting any published stratigraphic range. It is also of course important that autapomorphies for terminal taxa are not omitted from any cladistic analysis, as this will alter how such taxa are interpreted when constructing phylogen-etic trees.

*Range extensions, Lazarus taxa, pseudoextinction,
and ghost lineages*

Having dealt with the practicalities of tree construction some ter-minology needs to be clarified (Fig. 6.5).

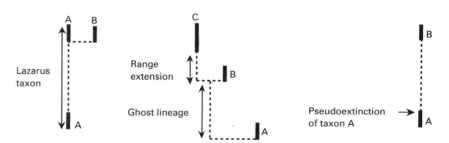

Fig. 6.5 Some terminology applied to phylogenetic trees.

A *range extension* is the extra stratigraphic range that must be added to the observed range of a taxon in order to construct an evolutionary tree concordant with the phylogenetic relationships deduced from morphological analysis. The observed stratigraphic range can be extended at either end. So, for example, in the case of a single hypothesized lineage linking two taxa, the earlier phenon can have its range extended up to the base of the derived phenon, or the later phenon can have its range extended back to the last occurrence of the earlier phenon, or both can have their ranges extended so as to meet midway. Since range extensions are all *ad hoc* assumptions about preservation or collection failure in the fossil record, phylogenetic trees ought to be constructed in a way that minimizes the range extensions that must be hypothesized.

A *Lazarus taxon* is a taxon that has a significant gap in its fossil record (Fig. 6.5). Here the taxon is recorded in the early and later part of its total range but is absent from the middle portion. This term was introduced by Flessa & Jablonski (1983) to refer to taxa that disappeared at mass extinction levels but which later reappeared (see also Jablonski, 1986b; Raup, 1986, 1987a; Smith, 1988a). Lazarus taxa disappear because of sampling failure, local extinction followed by recolonization, or simply because the appropriate facies are not developed at those levels in the sections studied. In its original formulation a Lazarus taxon undergoes no discernible morphological change deemed significant by a taxonomist between its first and last appearance. In such cases it is obvious that the gap in the taxon range is due to problems associated with the fossil record.

However, if the gap is sizeable, it is unrealistic to expect all taxa to remain morphologically unchanged, and new derived characteristics are likely to be acquired during this interval. There will then be two named phena separated by a gap, the older entirely plesiomorphic with respect to the later (Fig. 6.5). This leads to the disappearance of a taxon in name only, which is termed *pseudoextinction* (Stanley, 1979). Pseudoextinction is a common problem for paraphyletic groupings, even at 'species level'. In Fig. 6.5, phenon A is stratigraphically older than phenon B and, if plesiomorphic with respect to B then the simplest assumption is that A has transformed into B through the acquisition of additional derived character states. Although phenon A disappears from the stratigraphic record, its disappearance is in name only, the result of taxonomic convention. Phenon A is defined and recognized by the occurrence of derived states, e.g. a' and b', but these are also the states that identify taxon (A+B) as a monophyletic group, since both are also found in B. The monophyletic taxon defined by having character states a' and b' has therefore not become extinct.

Another way of looking at this problem is to consider the fate, not of taxa, but of derived characters. In extinction, the disappearance of a taxon is equivalent to the permanent loss of one or more derived character states (loss of genetic diversity). In contrast, the disappearance of a taxon through pseudoextinction is not accompanied by any loss of derived character states, only by the acquisition of one or more additional states (no loss of genetic diversity).

A *Ghost lineage* is an entire branch of an evolutionary tree for which there is no fossil record, but which needs to be hypothesized after combining cladistic and biostratigraphic data. The term was introduced and developed by Norell (1992). A ghost lineage will have a predicted set of morphological attributes, but its stratigraphic range is set by convention. For example, in Fig. 6.5 morphological analysis identifies B and C as sister taxa, and A as sister taxon to (B+C). Since A has its own unique derived characteristics it cannot be ancestral to (B+C). Therefore in constructing an evolutionary tree from these data it is necessary to hypothesize a common stem branch that runs from the dichotomy of B and C (at the first appearance of B in the stratigraphic record) to the first appearance of A. This branch is what Norell termed a ghost lineage. In this case, members of the ghost lineage will have all the derived morphological character states that are synapomorphies for the taxon (A+B+C) and may have one or more of the derived characters that are synapomorphies for the taxon (B+C). It will lack the derived characters that are unique to A, B, or C. The duration of this branch is very poorly constrained since both its upper and lower limits are defined by convention, i.e. the range extension is kept to a minimum.

Note that if A is a plesiomorphic phenon then there is no ghost lineage, since the resultant tree treats A as the ancestral phenon to (B+C) and involves only range extension.

Using phylogenetic trees to estimate
absolute taxonomic ranges

By constructing rigorous cladistic hypotheses of relationships and then using these to construct evolutionary trees on the basis of observed ranges of sister taxa, it is possible to infer where there are stratigraphic gaps in the fossil record. In contrast to the methods of graphic correlation or confidence interval calculation, sampling theory is unimportant to the phylogenetic method. Indeed range extensions can even be identified for taxa known from a single locality and horizon. Furthermore the phylogenetic method, unlike graphic correlation, allows range extrapolation outside the observed range of taxa. Finally, as first occurrences of monophyletic sister taxa each provides an estimate of their common time of divergence, taxonomic originations are likely to be more accurate than if the origin of each taxon was read directly from the fossil record. For these reasons, the taxonomic ranges deduced from phylogenetic trees are likely to provide truer approximations to absolute ranges than can be obtained by any other method.

The origin of angiosperms has been much debated and provides an excellent example of the benefits of applying a phylogenetic approach. The oldest fossils attributable to angiosperms are early Cretaceous, and a direct reading of the fossil record would imply that angiosperms evolved at around that time. However, the fossil record is not necessarily a reliable guide and Doyle & Donoghue (1993) examined the problem from a cladistic stance. They constructed a cladogram of seed plants which identified the immediate outgroups to the angiosperms. The sister group to angiosperms includes three clades with a long fossil record: the Gnetales, Bennettitales, and *Pentoxylon*. Together these three clades and the angiosperms form a monophyletic group, the Anthopytes, whose sister group is the Caytoniaceae.

The stratigraphic record of these groups, all of which are monophyletic, is shown in Fig. 6.6. When stratigraphic and cladistic data are combined to form an evolutionary tree, an estimate of the divergence time of angiosperms (node A) is obtained which is very much earlier than suggested by the fossil record. The range of angiosperms must be extended back to the first occurrence of its sister clade in the late Triassic. This estimate is strengthened from the fact that Gnetales, Bennettitales, and *Pentoxylon* all have fossil records extending back to the late Triassic or early Jurassic; the sister group is clearly early Mesozoic in origin. Doyle & Donoghue's phylogenetic approach provides evidence for a late Triassic origin of angiosperms, and em-

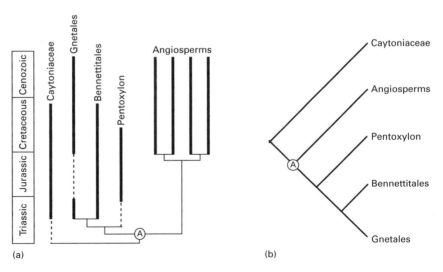

(a)   (b)

Fig. 6.6 (a) Tree and (b) cladogram for higher seed plants (Doyle & Donoghue, 1993). The phylogenetic tree constructed from the cladogram plus biostratigraphical evidence indicates that the origin of angiosperms (node A) must be at least as old as late Triassic even though the oldest fossil angiosperms are only early Cretaceous.

phasizes how the fossil record can seriously underestimate taxonomic ranges.

Examples of phylogenetic tree construction in practice

There are as yet relatively few well documented examples where morphological and stratigraphic data are presented as independent lines of evidence in the construction of trees. Archibold (1993) has given three examples from late Cretaceous–early Tertiary mammal faunas of North America. Here I draw four other examples from the literature. The first two deal with Mesozoic echinoids and early Palaeozoic trilobites and are relatively straightforward. The third involves Tertiary gastropods with relatively few characters, and exemplifies the approach needed where there are many unresolved polychotomies. The final example, a study of temnospondyl amphibian relationships, examines the problems encountered in the early stages of any analysis, where there are still many groups of uncertain phylogenetic status.

Arbacioid echinoids (Fig. 6.7 & Table 6.1)

Arbacioids are a group of regular echinoids found living today in habitats ranging from intertidal to deep-sea. Their fossil record extends back to the Jurassic. Smith & Wright (1993) provide a cladistic analysis for this group at generic level, based on a revision of representative species. Their original data matrix of 14 ingroup and 8 outgroup taxa has been reduced so as to include only two close outgroup taxa. There are 40 skeletal characters that can be scored in all taxa. Missing data form only about 1% of the matrix. Numerical analysis using PAUP (Swofford, 1993) found three equally parsimonious solutions 68 steps long (consistency index 0.66, retention index 0.76) (Fig. 6.7a–c). One of these solutions placed one of the outgroup taxa within the ingroup and can be rejected on the basis of the full analysis using all 8 outgroups. A combinable components consensus tree for the remaining two trees is shown in Fig. 6.7d and requires a single trichotomy.

A test run of 200 bootstrap replicates found that support for most branches was relatively high (Fig. 6.7d). The stratigraphic distribution of taxa is shown in Fig. 6.7e. The fossils attributed to each genus have been checked to ensure they have the correct defining characters and thus can validly be used to establish the first and last occurrence of each terminal taxon. Finally the branching pattern deduced in the cladogram is used to connect the observed biostratigraphic occurrence data to construct a phylogenetic tree. Some taxa, e.g. *Codiopsis*, are entirely plesiomorphic with respect to their sister taxa. In the cladogram, extinct taxa are indicated by an asterisk (*), taxa that are monophyletic are indicated by plus sign (+) after their name, and plesiomorphic metataxa are indicated by a minus sign (−). On the tree

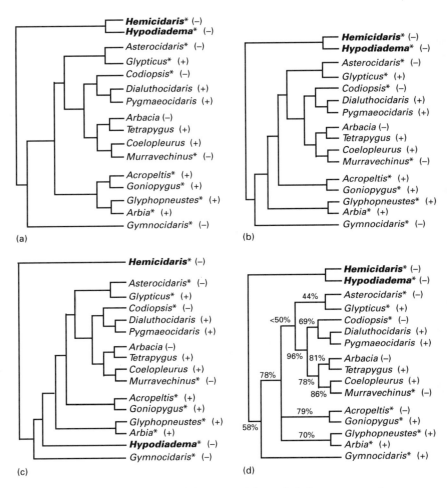

Fig. 6.7 (*Above and opposite.*) Cladogram and derived phylogenetic tree for arbacioid echinoids (Smith & Wright, 1993). (a–c) The three equally parsimonious cladograms found from a branch and bound parsimony analysis. Outgroup taxa are in bold. An asterisk* after the name of the terminal taxa indicates that the taxon is extinct; (−) indicates that there are no apomorphic states supporting this branch and that the group is therefore a metataxon; and (+) indicates that the taxon shows a derived apomorphic state and is thus monophyletic. (d) The combinable-components consensus tree for three cladograms, assuming that *Hypodiadema* is a member of the outgroup. Bootstrap support based on 200 replicates is indicated for each internal branch. (e) Stratigraphic distribution of arbacioid taxa (heavy lines) and inferred phylogenetic tree (dashed lines) based on cladogram (a).

the ranges of monophyletic taxa are shown as solid lines, those of metataxa as dashed lines. In cases where the derived sister taxon post-dates the plesiomorphic sister taxon it is simplest to assume that one arose from within the other and that the older metataxon is really a paraphyletic group. In the case of *Hypodiadema*, for example, few of the nominal phena assigned to this genus have ever been adequately

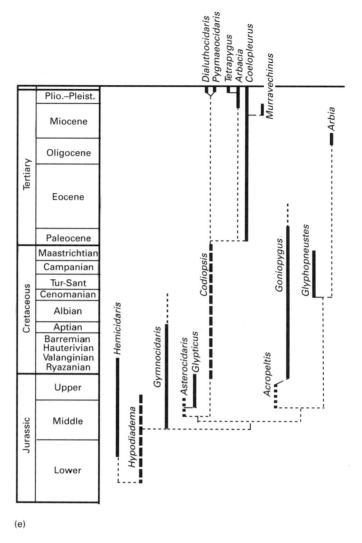

(e)

Fig. 6.7 *Continued.*

described and with revisionary work it will almost certainly be possible to split up this taxon.

Some range extensions are required, for example that linking the recent deep-water *Dialuthocidaris* and *Pygmaeocidaris* to the late Cretaceous *Codiopsis*. Several ghost lineages are also necessary, for example to extend the stem group of *Arbia* and *Glyphopneustes* back to the late Jurassic. The Oligocene *Arbia* shows the greatest discordance between stratigraphic and cladistic data; it is sister group to the late Cretaceous *Glyphopneustes*, sharing with that taxon derived features of the apical disc and peristome. The branch is reasonably well supported in the bootstrap analysis. However, *Glyphopneustes* has its

Table 6.1 Character matrix for 14 taxa of arbacioid and two hemicidaroid echinoids (full data available in Smith & Wright, 1993). Outgroup taxa are indicated by an asterisk *

Input data matrix

Taxon	1	2	3	4	5	6	7	8	9	1 0	1 1	1 2	1 3	1 4	1 5	1 6	1 7	1 8	1 9	2 0	2 1	2 2	2 3	2 4	2 5	2 6	2 7	2 8	2 9	3 0	3 1	3 2	3 3	3 4	3 5	3 6	3 7	3 8	3 9	4 0
*Hemicidaris**	0	0	0	1	0	0	0	0	1	0	0	0	0	0	1	0	0	0	0	0	0	0	0	1	0	0	2	1	1	0	0	2	0	0	0	0	0	0	0	0
*Hypodiadema**	1	0	0	0	0	1	0	0	0	0	0	0	0	0	0	0	0	0	0	0	0	0	0	0	2	0	0	0	0	1	0	0	0	0	0	0	0	0	0	0
Asterocidaris	1	0	0	0	1	0	0	0	0	1	1	1	1	0	1	0	0	0	0	0	1	0	0	0	1	0	2	0	0	0	0	1	1	0	0	1	0	0	0	0
Gymnocidaris	1	0	0	0	0	1	0	0	0	0	1	0	0	0	1	0	0	0	0	0	1	0	0	0	1	0	2	0	0	0	0	2	0	0	0	0	0	0	0	0
Glypticus	1	0	1	0	1	0	0	0	0	1	1	1	1	0	0	0	0	0	1	1	1	0	0	1	0	1	0	0	0	1	1	0	0	1	0	0	1	0	0	0
Codiopsis	1	0	0	0	1	0	0	0	0	0	1	1	1	0	1	0	0	0	0	2	1	1	0	0	1	1	0	0	0	1	1	0	0	1	0	0	1	1	0	0
Dialuthocidar	1	?	0	0	1	0	0	0	0	0	1	1	1	0	1	0	0	0	0	2	1	1	0	0	1	1	0	0	0	1	1	0	0	1	0	0	1	?	0	0
Pygmaeocidaris	1	2	0	0	1	0	0	0	0	0	1	1	1	0	1	0	0	0	0	2	1	1	0	0	1	1	0	0	0	1	1	0	0	1	0	0	1	?	0	0
Acropeltis	1	1	1	0	2	0	0	1	0	0	1	1	1	0	0	0	0	0	0	?	1	1	0	0	0	0	0	0	0	1	0	1	1	0	0	0	0	0	0	0
Goniopygus	1	1	1	0	2	0	0	1	0	0	1	1	1	0	0	0	0	0	1	1	1	0	0	0	0	1	0	0	0	0	1	0	2	0	0	0	0	0	0	0
Glyphopneustes	1	1	0	0	0	1	1	0	0	0	1	1	1	0	0	0	0	0	0	1	1	1	0	0	0	0	2	0	0	0	0	1	0	2	0	1	0	0	0	0
Arbia	1	0	0	0	0	1	1	0	?	0	1	1	1	0	0	0	0	0	0	?	1	1	0	0	0	1	2	0	0	1	0	1	0	1	2	1	0	0	0	0
Arbacia	1	0	0	0	1	0	0	0	0	1	1	1	0	0	0	0	0	0	2	1	1	1	0	0	1	0	0	0	1	1	0	1	1	1	0	0	1	1	1	
Tetrapygus	1	0	0	0	1	0	0	0	0	0	1	1	1	0	0	0	0	1	0	2	1	1	1	0	0	1	0	0	0	0	1	0	1	1	1	0	0	1	1	?
Coelopleurus	1	0	0	0	1	0	0	0	0	0	1	1	1	0	1	0	0	0	1	2	1	1	1	0	1	0	0	0	0	1	1	0	1	1	0	1	0	2	1	1
Murravechinus	1	0	0	0	1	0	0	0	0	0	1	1	1	0	1	0	1	0	0	0	1	2	1	1	1	0	1	1	0	0	0	1	1	0	1	1	0	1	0	1

own set of apomorphies and is unlikely to be directly ancestral to *Arbia*. The lineage of *Arbia* thus must be traced back to the base of *Glyphopneustes*.

The trichotomy in this particular case appears to be the result of a lack of phylogenetically informative characters rather than a conflict of characters. Nevertheless, one of the two cladograms has been chosen as a base for the phylogenetic tree. As it happens, the resolution of this trichotomy is largely irrelevant to the resultant tree, since the two cladograms imply more or less the same amount of range extension.

Asaphine trilobites (Fig. 6.8)

Fortey & Chatterton (1988) provided both a numerical cladistic analysis and a more subjective 'weighted' analysis (using only characters considered to be particularly important) for 10 families or superfamilies of the Cambrian–Ordovician trilobite suborder Asaphina. They used 42 characters in total, mostly discrete, and clearly indicated, for variable characters, how trait partitioning had been made. They found two equally parsimonious solutions and constructed an Adams consensus tree; my reanalysis of their data found three equally parsimonious solutions, but the same Adams consensus tree. Fortey & Chatterton provided only a sketchy evolutionary tree based on their cladistic analysis and the known stratigraphic ranges of the taxa.

Edgecombe (1992) also reanalysed Fortey & Chatterton's data, re-

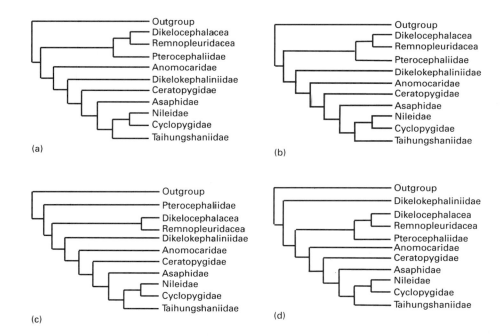

(a)

(b)

(c)

(d)

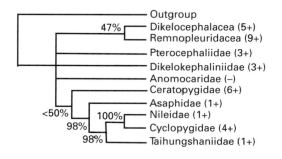

(e)

(f)

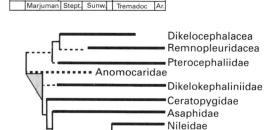

(g)

Fig. 6.8 Cladograms and evolutionary trees for the early Palaeozoic trilobite clade Asaphina (Fortey & Chatterton, 1988; Edgecombe, 1992). (a–d) The four equally parsimonious cladograms derived from the data matrix as modified by Edgecombe. (e) The strict consensus tree derived from cladograms (a–d). (−) indicates that terminal branch is unsupported by any apomorphy and is thus a metataxon; and (5+) indicates terminal branch supported by five apomorphies (numbers vary). Bootstrap support percentages are indicated for internal branches. (f) The phylogenetic tree constructed by Edgecombe (1992) from the consensus cladogram. (g) The phylogenetic tree constructed using the most parsimonious cladogram that corresponds most closely to the stratigraphic evidence (cladogram (a)).

moving two characters of some uncertainty. This revised data set gives four equally parsimonious solutions (Fig. 6.8a–d) and a combinable components consensus tree for these four solutions is given in Fig. 6.8e. Two hundred bootstrap replicates give excellent support for four of the branches, and moderate support for a fifth (Fig. 6.8e), but none of the remaining branches is supported and thus they must be treated with caution. Edgecombe then constructed a phylogenetic tree combining stratigraphic data with cladogram topology (Fig. 6.8f). However, in doing this he used a consensus tree as a basis for his phylogenetic tree and he treated all terminal taxa as if they were monophyletic. This resulted in an inaccurately reconstructed tree.

The consensus topology in this case results from conflict in the data set due, to a large extent, to the unstable position of Dikelokephalinidae in the hierarchy. The consensus tree indicates only those parts of the tree where branching pattern is uncontradicted and *does not* in this case correspond to any of the four most parsimonious solutions. Specifically, the basal polychotomy here does not indicate a multiple early divergence of taxa. It is therefore wrong to use the consensus topology as if it were a most parsimonious solution, since this will often imply range extensions when none is needed. As an alternative I have selected one of the topologies on the basis of stratigraphic evidence, choosing the one that requires the fewest *ad hoc* range extensions (Fig. 6.8a).

The second error made by Edgecombe was to treat all taxa as if they were monophyletic. That this is wrong can be seen by contrasting two taxa, Anomocaridae and Asaphidae. The Anomocaridae is entirely plesiomorphic in Fortey & Chatterton's data set and thus represents a metataxon that will certainly prove to be paraphyletic on closer inspection. This means that it will probably break down into a series of nested taxa each of which is successively more closely related to other more derived members of the clade. There is therefore no justification for extending ranges of more derived taxa back to the base of the Anomocaridae, and the fewest *ad hoc* range extensions are implied by deriving subsequent groups from the Anomocaridae when each first appears in the fossil record (Fig. 6.8g). Thus, whereas Edgecombe draws three range extensions and one ghost lineage back to the first appearance of Anomocaridae, none of these extrapolations is justified.

The opposite situation is found with the Asaphidae. This family is characterized by having a very distinctive supramarginal dorsal glabellar suture, a derived trait that marks the group as monophyletic. Apart from the earliest possible member (*Griphasaphus griphus* from the latest Middle Cambrian of Australia which is inadequately known and therefore only tentatively included), all Asaphidae show this trait. Although secondary loss is possible, the assumption must be that more derived sister taxa (Taihungshaniidae, Nileidae, and Cyclopygidae), which do not appear until the start of the Ordovician, must have split from the Asaphidae prior to their having evolved this derived trait. Therefore it is necessary to construct a ghost lineage extending from the first definitive Asaphidae (early Upper Cambrian) to the first appearance of a member of the derived clade (Taihung-shaniidae, in the earliest Tremadocian) (Fig. 6.8g).

The above example clearly demonstrates the importance of determining the status of the terminal taxa used in a cladistic analysis prior to tree construction. It also demonstrates the error of mistaking a consensus tree for a cladogram and how this can lead to overestimation of range extensions.

Ancilline gastropods

Michaux (1989) undertook a detailed analysis of the subfamily Ancillinae, a group of marine gastropods that is relatively widespread and common today in the Indo-Pacific region. He set out to resolve relationships for New Zealand Tertiary and Recent phena assigned to the genus *Amalda*. This genus is one of the largest of the group and has previously been subdivided into at least six subgenera.

Michaux's study encompassed 30 phena, assigned to five subgenera, with two outgroups. He was interested in integrating fossil and Recent phena, therefore he compiled a data matrix from shell characteristics alone. Gastropods have relatively few skeletal characters that can be coded in comparison to vertebrates, echinoderms, or arthropods, but Michaux was able to identify 17 characters, of which five were simple binary presence/absence characters, nine were subjective assessments of shell-shape features, one was a continuous variable ratio, and two came from discrete coding of clusters derived from multivariate analysis based on overall shell shape and size. A few phena turned out to have identical scoring and thus must be distinguished on features other than shell morphology. These have been combined for the present analysis.

Reanalysis of Michaux's data did not reproduce his findings exactly. Specifically, I could find no support for treating either *Baryspira* or *Gracilispira* as monophyletic. However, because of the relatively few characters available, a large number of equally parsimonious trees were produced from the data, and both Michaux's and my own reanalysis had to use heuristic search methods which do not guarantee finding the most parsimonious solution. The data do provide reasonable resolution for some parts of the cladogram (Fig. 6.9a). There is a basal dichotomy between phena assigned to *Alocospira* and to other taxa (*Amalda sensu stricto*). Among the latter clade, two phena of *Gracilispira* – *G. morgani* and *G. gigartoides* – are sister group to the rest. Finally, members of *Gemaspira/Spinaspira* form a well defined clade, supported by at least four derived character states. The remaining phena form a polychotomy which appears to be created solely by a lack of informative characters. In the cladogram, phena that are mono-

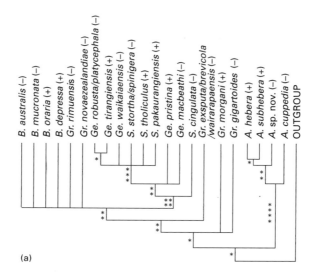

(a)

(b)

Fig. 6.9 Cladogram (a) and phylogenetic tree (b) for Tertiary ancillinid gastropods from Australia and New Zealand (Michaux, 1989). The cladogram shown is a combinable components consensus based on 34 equally parsimonious trees, with a consistency index of 0.61 and a retention index of 0.85. Taxa involved in polychotomies were omitted from tree construction. *, apomorphy supporting internal branch; (−), terminal phenon not supported by any apomorphy and thus a metataxon; (+), terminal taxon supported by one or more apomorphies. Observed stratigraphic ranges are shown as thick bars in the phylogenetic tree; dashed lines show the inferred phylogenetic tree.

phyletic are indicated by a plus sign (+) after their name, whereas those that represent plesiomorphic metataxa by a minus sign (−).

Michaux provided stratigraphic ranges for most of the taxa (Fig. 6.9b). *Gracilispira rimuensis, G. novaezealandiae,* and the *Baryspira phena* are unresolved in the cladogram, therefore it is not possible to generate a fully resolved evolutionary tree from these data and taxa involved in the polychotomy have been omitted. However, many of the branches can be reconstructed unambiguously and others have been identified from the majority rule tree. The basal dichotomy in *Alocospira* is between *A. cuppedia* and *A.* sp. nov. *A.* sp. nov. is stratigraphically the oldest, so it is tempting to place it as ancestral to the others. However, *A.* sp. nov. has a derived spire shape in common with *A. hebra,* whereas *A. cuppedia* retains the outgroup state. This evidence suggests that *A.* sp. nov. cannot be directly ancestral to *A. cuppedia,* and that the range of *A. cuppedia* should be extended back to the first occurrence of *A.* sp. nov.

The oldest *Amalda* species is *Gracilespira morgani* from the Upper Eocene. Since it is also the sister group to all other *Amalda* spp. it might be taken as a possible ancestor for the group. However, *G. morgani* also possesses an autapomorphy: an unusually wide, depressed band on the shell. This is derived by outgroup comparison, both with *Alocospira* and with more distant outgroup taxa; again, the oldest phenon cannot be considered as directly ancestral to any of the later members of this clade. *G. gigartoides,* on the other hand, is a metataxon, without any autapomorphies. It is therefore potentially paraphyletic and could contain members that have given rise to other phena. Its immediate sister taxon in the derived sister clade is *G. brevicola–G. wairapaensis–G. exsputa* complex (a single phenon in the morphological analysis), which occurs in virtual stratigraphic continuity. However, this complex is not stratigraphically the oldest member of that clade. Thus the available morphological evidence is also against taking *G. brevicola* as a direct descendant of *G. gigartoides.*

The available morphological evidence from shell form for resolving the phylogenetic relationships of *Amalda* is unfortunately not as strong or decisive as one would hope. However, it does provide some resolution and offers an important check on tree construction. Traditionally, the fossil record has simply been accepted uncritically as a correct record for fine resolution of phylogenetic relationships. With morphological data it is possible to check these assumptions and suggest where stratigraphic evidence is inadequate.

Temnospondyl amphibians (Fig. 6.10)

The temnospondyls are the largest group of archaic amphibians, and are currently classified in about 40 families and 160 genera, ranging from the Carboniferous to early Cretaceous. Milner (1990) revised the taxonomic relationships of this group at family level. His study is far

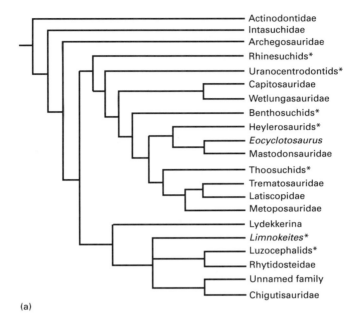

(a)

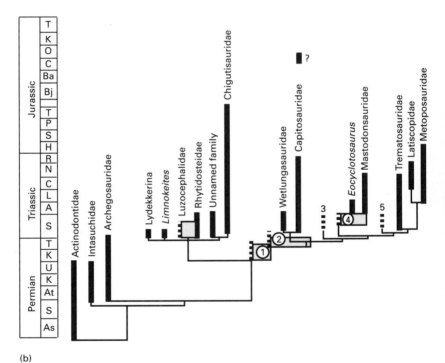

(b)

Fig. 6.10 Cladogram (a) and phylogenetic tree (b) for selected temnospondyl tetrapods (Milner, 1990). Grade taxa are shown on the cladogram by asterisks*, and on the phylogenetic tree as dashed bold lines. The stippled boxes indicate the possible geological period over which derived groups may have arisen. Unresolved grade taxa (metataxa) are numbered as follows: 1, rhinesuchids; 2, uranocentrodontids; 3, benthosuchids; 4, heylerosaurids; 5, thoosuchids.

from ideal, since he did not provide a data matrix and noted that his apomorphy list 'omits various reversals and variations of character states that occur within families'. Thus, there is no direct means by which others can assess the consistency of the characters stated to be diagnostic. Nor is the analysis subject to test through numerical cladistics. However, it does demonstrate the problems of plesiomorphic higher taxa.

In view of the preliminary nature of the work and the lack of revisionary studies in many groups, Milner had to accept that a number of his taxa were not demonstrably monophyletic and thus represented grades (metataxa) rather than clades. Milner's problem is common-place: given our current lack of revisionary systematics, many higher taxa used as terminal units in a cladistic analysis inevitably turn out to be non-monophyletic; yet to omit them from analysis because they are not demonstrably monophyletic may lead to the wrong topology. Furthermore, their stratigraphic distribution can often help define the first appearance of the larger monophyletic group to which they belong. But a start always has to be made somewhere, and cladistic analysis is not dependent upon the complete phylogenetic resolution of lower-level taxa. Systematics progresses through reciprocal illumination from many levels in the hierarchy. High-level metataxa will almost certainly turn out to be paraphyletic, but they are still useful during the early stages of the cladistic analysis of any group. Thus an initial higher-level taxonomic study that identifies metataxa and mono-phyletic taxa will help to identify the most appropriate outgroups for subsequent work aimed at resolving relationships within metataxa. Confusion should not arise, provided metataxa are recognized for what they are, namely groups awaiting more highly resolved analysis, not ancestors. Until significantly more systematic effort is mustered, metataxa are unavoidable in most analyses.

Milner constructed his cladogram placing most plesiomorphic taxa, not at termini, but as named grades on internal branches (although he was not entirely consistent in this). In Fig. 6.10a, all taxa, whether demonstrably monophyletic or not, are placed at the ends of branches, but with metataxa clearly distinguished from monophyletic taxa. Milner also provided detailed stratigraphic data for all taxa and constructed a phylogenetic tree along the lines outlined above. Part of his analysis is shown in Fig. 6.10b and illustrates several of the points discussed in previous sections.

The branching pattern from the cladogram defines the way in which stratigraphic records are connected. Thus it is able to predict gaps in the fossil record. The Trematosauridae are identified as sister group to (Latiscopidae plus Metoposauridae). The two groups are monophyletic, each supported by a number of derived characters, and therefore they must have diverged prior to the first appearance of either. Trematosauridae largely predates its sister taxon, therefore the stem group of (Latiscopidae plus Metoposauridae) must be extended back to the first occurrence of a member of the Trematosauridae. This almost doubles the range of the taxon (Latiscopidae plus Metopsauridae). There are two other significant range extensions that must be inferred. Firstly the clade formed by Intrasuchidae and all taxa to its right must be extended back to the first appearance of the monophyletic Actinodontidae. Secondly the clade of rhinesuchids plus its sister group must be extended back to the base of the Archegosauridae. On the whole, however, there is a reasonably good match between stratigraphic order and cladistic rank.

Plesiomorphic taxa fall into two classes: those that predate their derived sister group (groups 1, 2, and 4 in Fig. 6.10b), and those that appear contemporaneously with or later than the first occurrence of their derived sister group (groups 3 and 5). In the first case, additional data and more detailed study will probably resolve the phena included into a nested hierarchical series leading to the derived sister group. The earliest stratigraphic appearance of a member of one of these plesiomorphic taxa sets the latest time of divergence for the higher taxa to which it belongs. Thus the appearance of metataxon 2 (uranocentrodontids) defines the latest divergence time for the clade (uranocentrodontids to Metoposauridae). Plesiomorphic taxa that appear stratigraphically contemporaneous or later than their derived sister group cannot contain phena that are putative ancestors to the derived sister group. However, they are not necessarily monophyletic since they may, on closer inspection, contain more than one clade. The appearance of a derived sister group early in the stratigraphic record (e.g. Trematosauridae) predicts that divergence of all clades deeper in the cladogram must predate this time.

Summary

The construction of phylogenetic trees involves the concept of ancestry either indirectly or through the identification of 'ancestral taxa'. However, ancestral taxa remain problematic in the fossil record. Even at phenon level, taxa do not give rise to other phena, rather they arise from differentiation at lower levels in the hierarchy. Furthermore, any taxon that gives rise to another taxon must be paraphyletic and thus the construct of taxonomic convention.

However, in practice all cladistic analyses involve a mixture of monophyletic taxa and operationally indivisible plesiomorphic taxa as their terminal units. The latter are grades and potentially ancestral, in the sense that they may contain members that gave rise to the derived sister group, and should be treated as such when constructing phylogenetic trees.

Tree construction involves three assumptions: (i) that the cladogram is robust and provides the best available evidence for the phylogenetic relationships of the taxa under study; (ii) that monophyletic taxa do not give rise to other monophyletic taxa through character reversal;

and (iii) that gaps in the stratigraphic record of taxa are minimized. Range extensions, which are required to make the biostratigraphic and phylogenetic evidence concordant, are *ad hoc* assumptions that have to be made about gaps in the fossil record. By adopting these protocols cladograms and biostratigraphic data can be combined to produce our best estimate of the tree of life. This then allows us to explore patterns of evolution through geological time.

7 Patterns from the fossil record

The fossil record provides us with our only access to historical patterns of biodiversity through geological time. It is from the fossil record that we infer periods when clades have diversified, giving rise to a range of new morphologies over a short period of time, or when they have suffered setbacks through extinction. However, the fossil record is notoriously unreliable and, taken at face value, may generate spurious patterns. The present chapter outlines why and how a phylogenetic approach offers the best hope for reconstructing evolutionary patterns accurately.

Why phylogenetic data are essential

In recent years there has been a surge of interest in using taxonomic data to investigate evolutionary patterns. This approach, termed the *taxic approach* (e.g. Levinton, 1988), treats each taxon at some specified level as an individual historical unit. Evolutionary patterns are then based on counts of taxonomic appearances and disappearances in the fossil record, and statistical methods are used to identify departures from randomness through time. The taxic approach requires accurate biostratigraphic information and a sound taxonomic framework, since species records have to be grouped into higher taxa in the first place. However, unlike tree-based approaches, it does not require detailed knowledge of the phylogenetic relationships of the taxa used.

The greatest strength of the taxic approach is that it is simple to apply. Large databases can therefore be compiled with relative ease, and the resulting patterns can be evaluated by a wide range of statistical tests. The 'Achilles heel' that seriously undermines this approach is its dependence on systematic practice. Throughout this book I have questioned the quality of traditional systematics and argued that non-monophyletic taxa are the product of taxonomic judgement, not biological processes. If we are to gain access to biologically meaningful patterns, we must base our observations on monophyletic taxa alone. As Fisher (1991, p. 106) stated, 'if analytical power is purchased at the price of the validity or relevancy of the fundamental units of analysis, we have to ask whether we are getting a good deal'.

Raup (1991) made the important point that databases need not be complete, nor need they be perfectly accurate: 'Random noise cannot, by itself, produce a significant signal where none exists' (Raup, 1991, p. 209). Yet there may be strong biases, relating to fluctuations in the quality of the geological record, that have led to distributions of non-monophyletic taxa departing from random (Smith & Patterson, 1988; Fortey, 1989). The important question, then, is How complete and how accurate must data be to be useful? This to some extent depends on the problem being tackled. If it is simply diversity patterns that are of interest, the nature of the taxonomic groupings used is largely irrelevant (see p. 88), but if extinction patterns are involved, the quality of the data can make a huge difference (e.g. Patterson & Smith, 1987; Archibold & Bryant, 1990; Archibold, 1993).

The few studies on this subject suggest that monophyletic taxa form only a small minority of traditional data (see p. 79), so the problem is not trivial. Furthermore, a number of taxic approaches rely on rank equivalency among taxa. Families, for example, are assumed to represent equivalent entities, and variation in the number of phena included is seen as having biological significance. But rank has meaning only within a single hierarchical scheme and can be only vaguely comparable among clades, due to the inconsistencies of taxonomic usage. The fact that families become richer in phena through time (Valentine, 1969; Sepkoski, 1984; Flessa & Jablonski, 1985) is more a comment on how taxonomists, as a community, have gone about their business than about evolutionary processes. Maybe taxonomists working in groups unconstrained by extant phena are more prone to generate 'families', or maybe better preservation and increasing organismic complexity towards the present have allowed taxonomists to make finer phenon-level discrimination. The possible explanations are many, but all relate to human intellectual endeavours, not biological processes.

The taxic method is crippled not only by the data it uses and the assumptions that must be made about taxa, but also by sampling problems. Consider the hypothetical example in Fig. 7.1, which shows the distribution of taxa A–M through four stratigraphic intervals. Conventionally, if a taxon is recorded anywhere within one stratigraphic interval, its range is extended to encompass the entire interval. For example, all taxa making their first appearance in interval 2 are assumed to have originated at the start of that interval, and a similar convention is used for extinctions.

Using the taxic approach, the standing diversity for each time interval is calculated by summing the number of taxa present. Thus intervals 1 and 3 have two taxa while intervals 2 and 4 have seven taxa each. The taxonomic origination rate is calculated from the number of new taxa appearing in each interval, often standardized against the total number of taxa present. Thus, five taxa originate in interval 2 (0.71), none in interval 3, and six in interval 4 (0.86) (originations in interval 1 cannot be calculated without data on the preceding interval). The same approach is used to calculate extinction rates, with no extinctions in interval 1, five in interval 2 (0.71), one in interval 3 (0.5) and an unknown number in interval 4 (since survival into the next interval is not tabulated). The taxic approach thus recognizes two

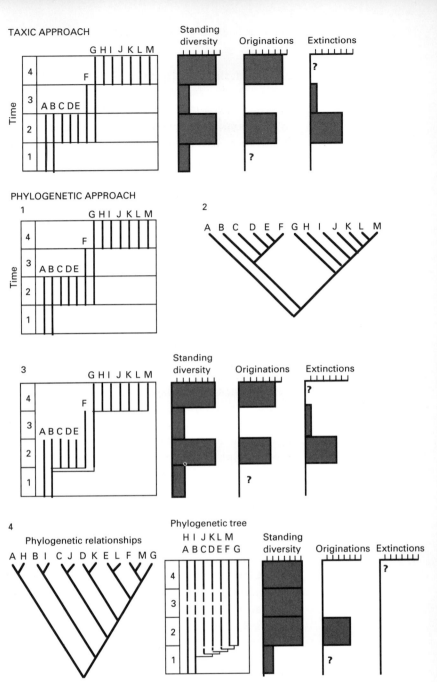

Fig. 7.1 Hypothetical scheme to demonstrate the importance of phylogenetic information in determining patterns of diversity, origination, and extinction. Given the same biostratigraphic data, very different patterns emerge depending upon the phylogenetic relationships deduced for the taxa. (a) Taxic approach. Biostratigraphic ranges of taxa are compiled and these are used directly to calculate standing diversity and rates of origination and extinction. (b–e) Phylogenetic approach. (L) Biostratigraphic ranges of taxa are compiled. (c) Phylogenetic relationships of these taxa are derived from cladistic analysis of character distributions. (d) Phylogenetic and biostratigraphic data are combined to produce an evolutionary tree. Standing diversity and rates of origination and extinction are derived from the tree topology. (e) Different cladistic hypotheses produce different phylogenetic trees and diversity patterns.

periods of high speciation (intervals 2 and 4) and a major extinction event (at the end of interval 2). Interval 3 marks a period of low diversity following the extinction event.

A very different picture can emerge using a phylogenetic approach. In addition to information on the biostratigraphic ranges of taxa, we also need to know how they are related historically. Consider two extreme examples. Firstly, if the cladogram identifies taxa B–F as one monophyletic group and taxa G–M as another (Fig. 7.1(2)), the evolutionary tree that makes the fewest *ad hoc* range extensions is as shown (Fig. 7.1(3)). This gives exactly the same results as the taxic approach in terms of standing diversity and origination and extinction patterns.

However, consider the second case, in which taxa H–M each has its plesiomorphic sister taxon among taxa A–F (Fig. 7.1(4)). When the evolutionary tree is constructed a very different pattern emerges. Instead of an extinction at the end of interval 2 followed by a period of low diversity and then a radiation in interval 4, interval 3 is identified as a period of preservation and/or collection failure. The extinction at the end of interval 2 disappears, as does the radiation at the start of period 4, and standing diversity remains constant through intervals 2–4.

Thus, a serious drawback of the taxic approach is that it lacks the power to discriminate between sampling patterns and genuine evolutionary patterns. Provided sampling biases are randomly distributed both among taxa and through time, the taxic approach will merely have an added level of noise to cope with, and the worst that can be expected is that a relatively weak signal will be missed. However, if preservational potential is in some way temporally biased, the taxic approach will do no more than track sampled diversity. For example, if variation in sea-level stand significantly alters the area of shelf over which there is net accumulation of sediment (as it surely must), then the chances of faunas being preserved and later discovered may fluctuate significantly through time. Under those circumstances the taxic approach will do no more than track sampling biases and a spurious signal may be generated. Without phylogenetic information the danger is that sampling patterns generated by variations in the quality of the fossil record will be mistaken for genuine evolutionary patterns.

If we are to stand the best chance of deciphering patterns in the fossil record that arise from biological phenomena, then we must adopt an approach that can overcome sampling bias and avoids using arbitrary taxa. Our best approach is to derive patterns from monophyletic groups alone and to eschew the use of taxonomic rank as any form of evolutionary yardstick. Patterns must come from analysis of phylogenetic tree structure, not taxonomic rank. Previous chapters have outlined how hypotheses of phylogenetic relationships can be constructed from morphological data, and how this can be integrated with biostratigraphic data to produce a phylogenetic tree. The structure of this tree provides our best means of assessing evolutionary patterns from the fossil record. Thus, periods of rapid branching, or congruence in the timing of branch initiation or termination on phylogenetic trees, represent patterns of biological significance. Individual trees alone will only give a piecemeal view of evolutionary processes but, by assembling data from a number of different clades over the same time interval, a nomothetic approach is possible. Patterns can be recognized that affect many clades simultaneously and whose causality is primarily extrinsic.

The phylogenetic approach to the study of evolutionary patterns in the fossil record is still in its infancy. Clearly the accuracy of the phylogenetic tree, and hence of the evolutionary patterns that we perceive, depends on the accuracy of our taxonomic and biostratigraphic data. This poses no serious problem, since both are being constantly improved and refined (e.g. Maxwell & Benton, 1990; Sepkoski, 1993). The remainder of this chapter explores the applications and advantages of the phylogenetic approach for documenting evolutionary patterns.

Biodiversity

There are many ways in which diversity can be measured. We could, for example, measure the number of phena present in any one interval. This would give us an indication of biological diversity: the number of biological species present. Another approach is to try to measure morphological diversity present in any one interval. This need not correspond to biological diversity because simple counting of phena does not consider the amount of morphological divergence that separates them. For example, the same number of species can occupy a small, well-defined area of morphospace densely or a larger area more sparsely (Foote, 1991a,b,c) (Fig. 7.2). These two kinds of diversity will be discussed separately.

Taxic diversity

The number of biological species existing at any one time can only be estimated from the diversity of phena in the fossil record. This will depart from true biological species numbers because of the well known problems associated with recognizing biological species from morphological data (e.g. the problems of sibling species and polymorphic species). However, if there is no systematic or temporal bias then we can expect phenon diversity to give a reasonably accurate measure of biological species diversity.

The record of phena is generally acknowledged to be so patchy (e.g. Raup & Boyajian, 1988; Valentine, 1990) that simple phenon counts through time are likely to be highly biased by the vagaries of the fossil

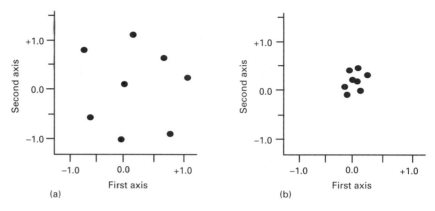

Fig. 7.2 The same number of taxa can occupy a relatively large area of 'morphospace' sparsely (a) or a small area densely (b). In this case morphospace is defined using the two principal component axes of a multivariate analysis.

record (Chapter 5). The taxic approach attempts to overcome sampling bias by using higher taxa as a proxy for estimating phenon-level diver-sity. Sampling biases are overcome through range interpolation between first and last occurrences of taxa in the stratigraphic record. This compensates for those intervals within the taxon's range where phena have not yet been recorded. The total diversity of a given time interval is thus constructed from the number of taxa that are estimated to have existed during each period, rather than from observed diversity. The taxic approach with range interpolation is an appropriate method of estimating diversity (p. 88), provided polyphyletic groups are not a major component of any database. The one drawback of this approach is that resolution must be sacrificed to compensate for preservational bias.

It is unnecessary to assume that rank has any biological significance in this context. When compilations of family-level data are used, there is no assumption that these represent real biological entities. The only assumption needed is that phena have been partitioned into monophyletic or paraphyletic groupings of very approximately similar size by taxonomists. As Raup & Boyajian (1988, p. 112) point out, 'we need not know precisely what a genus or family is as long as there is an operational correspondence between rises and falls in species, generic and familial patterns'.

One obvious difficulty is that no strict correlation exists between numbers of phena and assigned taxonomic rank, because rank has been established on an *ad hoc* basis. This problem can be overcome by using large databases. Averaging over enough data will effectively dampen the variation associated with taxonomic rank. On average, taxa of higher rank contain more phena, although this is simply a result of increasing variance and tree topological constraints.

Higher taxa, such as genera or families, can therefore give an estimate of the number of phena that have existed for periods in the geological past. The lower the taxonomic category used, the better it will be for estimating numbers of phena, since the variance associated with estimating the average number of phena included in a taxon will generally decrease down the hierarchy. The taxic approach removes some of the sampling bias by range interpolation between the first and last observed occurrences of a taxon. However, it does not allow extrapolation outside the observed range, and thus cannot identify periods when the taxon existed either before the first recorded level or after the last recorded level.

Other biases become important only when diversity patterns are compiled over reasonably long periods of time. Firstly there is an apparent temporal bias in the origination or extinction of the taxa being used; and secondly there is an apparent bias in the average numbers of phena assigned to higher taxa through time.

Temporal bias in the origination or extinction of taxa. The numbers of phena included in a higher taxon are not uniform through time. Each higher taxon should increase in diversity from a single phenon at the start of its range, and declines to one or a few phena at the end of its range. The way in which diversity of phena increases and decreases during the history of a clade varies from higher taxon to higher taxon. If taxonomic originations and extinctions are randomly distributed through time, then averaging over a large enough sample will compensate for these individual variations. However, this is clearly not the case since the distribution of originations and extinctions in the fossil record is demonstrably far from random (Sepkoski & Raup, 1986; McKinney & Oyen, 1989). Thus, for periods when there is a predominance of taxonomic originations the diversity of phena is likely to be overestimated.

Temporal bias in the number of phena assigned to taxa. The ratio of phena to families increases by a factor of two from the Mesozoic to the Cenozoic (Valentine, 1969; Sepkoski, 1984). Flessa & Jablonski (1985) also pointed out the general increase in numbers of phena assigned to families during the Phanerozoic as a whole. Since rank is a matter of convention, this phenomenon represents the vicissitudes of taxonomic practice and the bias, if uncorrected, will progressively overestimate the numbers of phena back through time.

The phylogenetic approach

The phylogenetic approach begins, like the taxic method, with recording the occurrence of phena in the stratigraphic record. The phylogenetic relationships of these phena are then established through formal morphological character analysis and a phylogenetic tree is constructed (as outlined in Chapter 6). This not only involves range interpolation between the first and last occurrence of a phenon, but can also identify where ranges need to be extended beyond that observed. Biodiversity of phena is then estimated from the numbers of observed and implied lineages on the tree at each time interval.

If higher taxa are used in place of phena, the method falters because of the additional assumptions involved. Higher taxa must be assumed to contain approximately equal numbers of phena on average if higher taxa are to be used as proxy to estimate phenon diversity. This implies that taxa of the same rank represent equivalent entities, which is demonstrably false. As in the taxic approach, only by averaging over a large sample can individual variation be compensated for.

The phylogenetic approach has clear advantages for relatively small studies in which data can be compiled at low taxonomic level. Firstly it provides a method for range extrapolation beyond the observed stratigraphic occurrence of a taxon. It can therefore compensate for sampling deficits even at phenon level. A second advantage is that it is not a probabilistic method and therefore does not require large sample sizes to be applied effectively.

However, for broad regional or global studies in which data must be compiled at relatively high taxonomic levels, both the taxic and phylogenetic methods estimate phenon diversity through extrapolation, and both suffer from the same problems. A large database is required because numbers of phena assigned to higher taxa are variable and can only be estimated on a probabilistic basis. This database is currently more easily obtained through a taxic approach and offers advantages over the phylogenetic approach.

Biodiversity of early Palaeozoic carpoids

An example of the phylogenetic method in practice is Cripps's (1991) analysis of cornute relationships. Cornutes are an extinct group of deuterostomes, variously regarded as chordates or echinoderms, that existed during the early Palaeozoic. Cripps provided a numerical cladistic analysis of the 34 described species of cornute, which range in age from Middle Cambrian to late Ordovician. He used three solutes (the primitive sister group to cornutes) for outgroup rooting and also included a composite coded taxon for the mitrates, another closely related group (Jefferies, 1986). A total of 75 characters were assembled, and the data matrix was presented together with the resultant clado-gram. An initial run with equal weighting to all characters found 68 equally parsimonious solutions, which provided partial resolution of relationships. Characters were then reweighted according to their consistency index, and the analysis rerun. This produced a solution that involved only two trichotomies.

Cripps also provided data on the stratigraphic occurrence of all his phena, although he did not construct a phylogenetic tree. The observed stratigraphic occurrence of the taxa is shown in Fig. 7.3a by solid circles, and the taxa are numbered. By following the branching pattern imposed by the cladogram and the temporal relationships imposed by

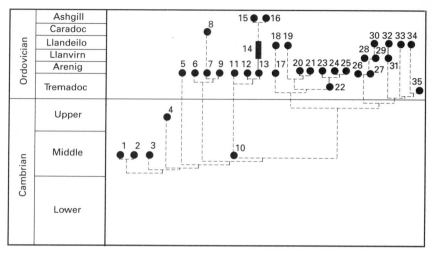

(a)

Fig. 7.3 (a) Phylogenetic tree constructed for cornutes (problematic deuterostomes) from the cladogram and stratigraphic data presented by Cripps (1991). Dashed lines indicate range extensions and ghost lineages that must be invoked to make the cladogram and stratigraphic data congruent. Taxa 1–35 are as follows: 1, *Ceratocystis perneri*; 2, *Ceratocystis vizcainoi*; 3, *Protocystites menevensis*; 4, *Nevadaecystis americana*; 5, *Cothurnocystis primaeva*; 6, *Cothurnocystis fellensis*; 7, *Procothurnocystis owensi*; 8, *Cothurnocystis courtessolei*; 9, *Cothurnocystis elizae*; 10, *Cothurnocystis bifida*; 11, *Thoralicystis melchiori*; 12, *Thoralicystis zagoraensis*; 13, *Thoralicystis griffei*; 14, *Bohemiaecystis bouckei*; 15, *Scotiaecystis collapsa*; 16, *Scotiaecystis curvata*; 17, *Amygdalotheca griffei*; 18, *Minolicystis kerfonei*; 19, *Phyllocystis salairica*; 20, *Phyllocystis blayaci*; 21, *Phyllocystis crassimarginata*; 22, *Prochauvellicystis semispinosa*; 23, *Chauvellicystis spinosa*; 24, *Chauvellicystis ubaghsi*; 25, *Chauvellicystis vizcainoi*; 26, *Galliaecystis lignieresi*; 27, *Progalliaecystis ubaghsi*; 28, *Hanusia prilepensis*; 29, *Hanusia sarkensis*; 30, *Hanusia obtusa*; 31, *Reticulocarpos hanusi*; 32, *Beryllia miranda*; 33, *Domfrontia pissotensis*; 34, *Prokopicystis merglii*; 35, mitrates.

the stratigraphic record, an evolutionary tree (shown by dashed lines) can be constructed. Minimal range extension has been invoked according to the methods set out in Chapter 6. For example, the observation that taxon 7 (*Cothurnocystis courtessolei*) is plesiomorphic with respect to its sister taxon 8 (*Cothurnocystis elizae*) and also precedes it stratigraphically, implies that it can be taken as a putative ancestor (note that virtually all phena are known only from single localities and horizons). When the stratigraphically early occurrence of a relatively derived taxon (e.g. taxon 35, mitrates) dictates that a series of more primitive sister groups (26–34) must have diverged during a period for which they have no record, topological constraints and nominal ghost lineages push these divergences back in time.

A direct count of the number of phena present through geological time indicates low diversity during the Upper Cambrian and Tremadocian, and highest diversity in the Arenig (Fig. 7.3b, solid bars). These data are raw, i.e. uncorrected for sampling error. Grouping the phena

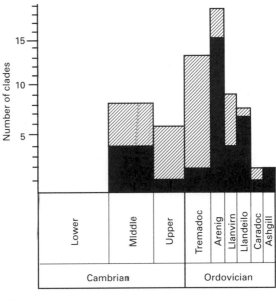

(b)

Fig. 7.3 *Continued.* (b) Estimated standing diversity for cornutes. The black bars indicate the observed taxic diversity through time, the stippled bars represent the additional diversity that is estimated to have been present when ghost ranges and range extensions are taken into account.

into genera does little to compensate for sampling deficits since there is a high proportion of monotypic genera, and those that are not monotypic are largely restricted to single stratigraphic horizons. Thus the generic diversity pattern follows the phenon pattern very closely. Again it indicates relatively low diversity through the Cambrian and Tremadocian, with a major peak in the Arenig. Grouping into taxa at family level is more difficult, because a number of these genera have never formally been placed in recognized families. However, plesions are set up by Cripps and, apart from two that are monotypic, these correspond approximately to traditional family groupings. These are too few to obtain a satisfactory plot of diversity over time.

When the evolutionary tree is used to calculate the numbers of clades present in each time interval, diversity in the Tremadocian increases tremendously. Furthermore, range extensions also add significantly to both the Upper and Middle Cambrian estimates. The tree suggests that, at the very most, only half of the Middle Cambrian lineages have been discovered, and more than 80% of the Tremadocian lineages remain to be discovered. Tree construction is done on the principle of minimal range extension, so the phylogenetic method will provide only a minimum estimate of lineage diversity. This highlights the fact that the low phenon count in the Tremadocian and Upper Cambrian almost certainly represents sampling failure.

By using the ratio of sampled to predicted lineages in phylogenetic trees such as this, we can obtain a measure of how good the fossil record is. The fossil record in the case of cornutes appears to be very poor during the Cambrian and Tremadocian, and best in the Llandeilo. Smith (1988a) used a similar approach for Cambrian echinoderms in general. Phylogenetic trees pinpoint where sampling deficiencies lie and, by compensating for them, produce more accurate depictions of lineage diversity than are possible through the taxic approach.

In summary, the taxic approach relies on higher taxa to dampen the problems of the patchy fossil record, and thus loses resolution. The phylogenetic method uses cladistic relationships to recognize and compensate for sampling/preservation failure without loss of resolution. The phylogenetic method thus has significant advantages at low taxonomic levels and for periods when there may be a temporal bias in the quality of the fossil record. At high taxonomic levels both methods rely on extrapolation and the variance associated with estimating numbers of phena from taxa increases significantly. This increased variance can only be compensated by using larger samples and the taxic approach then becomes more attractive for practical reasons.

Diversity profiles

Changes in the diversity of taxa are often depicted by means of spindle diagrams, and the shape of these diagrams has been investigated for its biological significance (Simpson, 1944; Gould *et al.*, 1977, 1987). This approach was formalized by Raup *et al.* (1973) who generated such 'clade' diversity diagrams by random simulation, as a null hypothesis for comparison with actual diversity diagrams. (Note that the 'clades' referred to by Raup and others are actually a mixture of paraphyletic and monophyletic groups: of the 14 'clades' illustrated as an example by Raup *et al.* (1973), only six are monophyletic.)

There has been much speculation as to the significance of 'clade' shape. Gould *et al.* (1977) used random models of 'clades' with constant speciation and extinction rates and found that many were remarkably like the patterns observed for major higher taxa in the fossil record. They speculated that chance alone could be sufficient to generate many evolutionary patterns. However, Gould *et al.* used unrealistic scaling; each clade was composed of only small numbers of phena. When the process was repeated using 'clades' of more realistic size, the similarity between model and observed pattern evaporated (Stanley *et al.*, 1981).

Another parameter that has been investigated is the relative position of a 'clade's' greatest diversity. 'Clades' may have their greatest diversity at the midpoint of their geological duration – in which case they are said to be symmetrical – or nearer the beginning (bottom-heavy) or end (top-heavy). Gould *et al.* (1987) looked at the distribution of 'clade' asymmetry through time and discovered that bottom-heavy 'clades' tended to predominate in the early stages of diversification of a major taxon. This was true both when comparing (at the level of genera within families) 'clades' originating in the Cambro-Ordovician as opposed to those arising later, and also for individual 'clades' such as Tertiary mammals (Fig. 7.4). This general pattern was viewed as biologically significant.

However, the claimed deviations of Cambrian and Ordovician clades towards bottom-heaviness were not statistically significant when tested (Kitchell & MacLeod, 1988). Furthermore, the 'clades' of Gould *et al.* (1987) encompass a sizeable proportion of paraphyletic groups. Paraphyletic groups are what is left over after abstraction of derived groups within a larger clade, and will therefore naturally predominate in traditional taxonomies at early periods in the history of any major taxon. Fortey (1989), for example, has shown that graptolites have a preponderance of paraphyletic taxa early in their history; the same is true for echinoderms (Smith, 1988a). Such paraphyletic groups are bottom-heavy simply because the more derived portions have been removed as monophyletic clades. The paraphyletic echinoderm class Eocrinoidea, which represents a group that is confined to the early part of the phylum's history, is clearly bottom-heavy (Fig. 7.5). However, this diversity profile is pure artefact, the result of taxonomic practice which has pruned off a series of derived groups. When eocrinoids are placed into a hierarchical scheme of monophyletic clades a very different pattern emerges (Fig. 7.5). Thus paraphyletic taxa may be expected to be bottom-heavy because they are truncated by taxonomic convention (see p. 76).

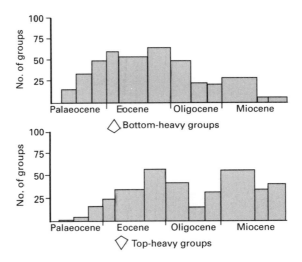

Fig. 7.4 Diversity histograms for top-heavy and bottom-heavy 'clades' (traditional taxonomic groups) for mammals derived by counting genera within families at each stage of the Tertiary (Gould *et al.*, 1987). Bottom-heavy groups tend to predominate early in the history of diversification.

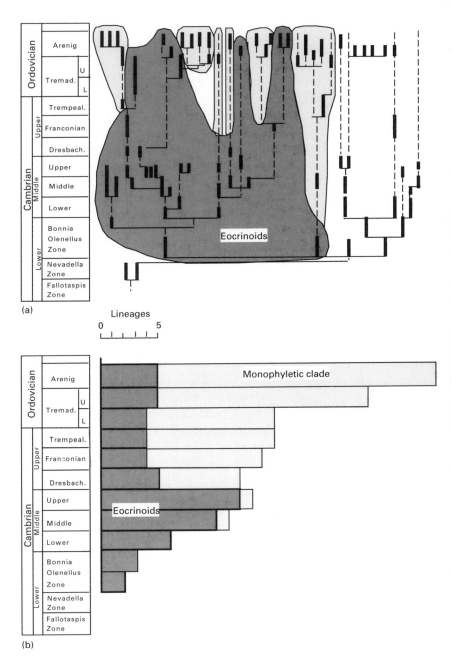

Fig. 7.5 (a) Phylogenetic tree for phena of early Palaeozoic echinoderms (from Smith, 1982, where full supporting data are presented). Phena assigned to the eocrinoids, a traditional, bottom-heavy group and a member of Sepkoski's (1984) Cambrian fauna, are represented by the densely shaded area. The group Eocrinoidea is paraphyletic and clades derived from the eocrinoids are indicated by lighter shading. (b) The diversity histogram for eocrinoids (black bars) is bottom-heavy. However, if the monophyletic clade of eocrinoids plus all of its descendants is considered (black bars plus stippled bars) the artificiality of bottom-heaviness in paraphyletic clades is demonstrated.

Even for monophyletic groups, the major problem with spindle diagrams is that they provide no insight into how diversification and extinction have interacted to generate any given profile. A simple spindle diagram which is 'fat in the middle and thin at both ends' can be produced by a variety of very different tree topologies. A proper understanding of diversity profiles can only come from the analysis of tree branching patterns and is therefore dependent on systematics.

Morphological disparity

Counting species is not the only way of measuring diversity; morphological disparity is equally important. We might want to know not only how many species existed at any time, but also what area of morphospace they occupied. Gould (1991) identified this as one of the future key areas for palaeobiological research.

Morphometric approaches

Traditionally, taxonomic rank has been used as a proxy for morphological distance, with taxa of higher rank assumed to encapsulate more morphological difference than taxa of lower rank. However, since taxonomic rank has been used primarily to indicate only relative inclusiveness, high rank does not imply that the earliest members of a taxon are separated from other taxa by large morphological distances. Rank is a notoriously poor indicator of morphological disparity (Foote, 1991a,b,c; Gould, 1991; Strathmann, 1991).

To overcome some of these problems quantitative morphometric techniques, such as principal component analysis and Fourier analysis of landmarks, have been applied to study the distribution of phena in multidimensional morphospace (Temple, 1980, 1987; David & Laurin, 1989; Laurin & David, 1990a,b; Tabachnick & Bookstein, 1990; Foote, 1991a,b; Rohlf & Marcus, 1993). This approach requires little or no a priori phylogenetic information about the taxa selected. However, any historical assessment of changing morphological disparity needs to be placed in a phylogenetic context.

Landmark techniques and other methods for the quantitative description of form are only really applicable to easily homologized structures, e.g. trilobite glabella or blastoid calyx. Nevertheless they have been used extremely successfully to document the evolution of morphological diversity within clades (Cherry *et al.*, 1982; Foote, 1991a,b, 1992a,b). For example, Foote (1991a, 1992b) assembled data on trilobite cranidial form using 12 Fourier coefficients to describe cranidial outline. He then compared the relative morphological variation of cranidial shape using principal component analysis at different periods in the geological past. As the technique is highly susceptible to sample-size factors (sample size and morphological diversity covary), and the trilobite samples were not equivalent in size, Foote used rare-faction to extrapolate morphological diversity to comparable levels. He was able to demonstrate that Middle and Upper Cambrian trilobites occupy a relatively small and well defined area of morphospace and display many variations on relatively few morphological themes. In contrast, Middle and Upper Ordovician trilobites occupy a larger range of morphospace at similar phenon densities (Fig. 7.6), at least in terms of cranidial outline. This study and others (e.g. Foote, 1992a) emphasize the point that extrapolating morphological diversity from taxonomic data can be prone to significant error.

The major drawback to morphometric comparisons is that they can only be applied successfully to easily recognized (and thus conserved) homologous structures. This not only limits their utility but may, by concentrating on relatively stable morphological features, ignore others more variable in form, giving a false indication of morphological disparity. Several workers have therefore recently turned to discrete character data in an attempt to document morphological distance at higher taxonomic levels (Foote, 1991c, 1992a; Briggs *et al.*, 1992; Smith *et al.*, 1992). A battery of phenetic techniques for defining and differentiating phena in morphospace can now be applied to questions of morphological disparity at higher levels. Briggs *et al.* (1992), for example, used principal component analysis to investigate the question of early arthropod disparity. Such techniques are, however, highly sensitive to sampling bias (Foote, 1992a; Foote & Gould, 1992).

Somewhat less sensitive to sample size is the technique of pairwise comparison (Foote, 1992a,b), in which the mean morphological distance separating all pairs of taxa at a given time interval is calculated. Pairwise dissimilarity can be calculated in one of two ways: either directly from a distance matrix (the phenetic method), or from character distribution on a cladogram (the phylogenetic method). Whereas the phenetic method calculates dissimilarity on the observed differences between two taxa divided by the total number of characters, the phylogenetic method calculates dissimilarity on the total number of morphological changes inferred to have occurred between two taxa from the time when they shared a common ancestor.

Foote (1992a, p. 7325) criticized morphological (cladistic) analyses that ignored plesiomorphic states and argued that quantitative (phenetic) methods were preferable because they took account of both primitive and derived aspects of morphology. The example given in Fig. 7.7 demonstrates that this is not necessarily the case.

Assume that the morphological character–taxon matrix for these five taxa is an unbiased assessment of the observed morphological differences amongst them. Distance matrix and parsimony analyses will both identify the same tree, which incorporates a small amount of homoplasy, affecting taxon C (Consistency index = 0.86). Pairwise distances calculated by counting the total number of observed differences between each pair of taxa are shown in the lower half of the

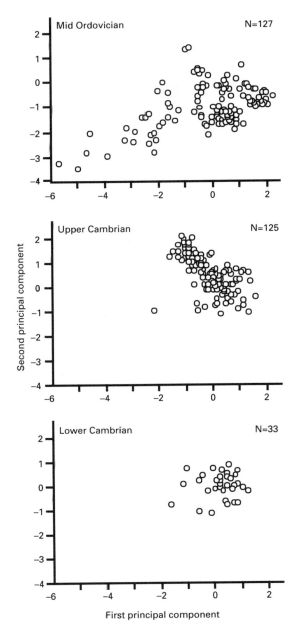

Fig. 7.6 Plots of the first two principal component axes from a multivariate analysis of trilobite cranidia. Cranidial shape was determined using perimeter-based Fourier analysis utilizing 12 coefficients. Each circle represents a single specimen but not necessarily a separate phenon since sampling was random. The axes are all to the same scale.

similarity matrix. As the distance method is based on comparison of terminal taxa alone, it takes no account of character convergence in the tree, mistaking homoplasy for homology. The parsimony method of pairwise comparison uses the same data matrix to construct the tree

Matrix
A 0000000111000110000011111110011
B 0000000110000001100011111101011
C 0000000000011100001111111100100
D 0001111111111000000000000011101
E 1110000111111000000000000011110

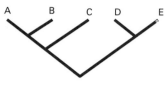

(a)

Absolute distances
Above the diagonal = cladistic pair distances
Below the diagonal = distance matrix pair distances

	A	B	C	D	E
A	–	0.22	**0.55**	0.58	0.55
B	0.22	–	**0.58**	0.61	0.58
C	**0.48**	**0.45**	–	**0.68**	**0.65**
D	0.58	0.61	**0.58**	–	0.29
E	0.55	0.58	**0.58**	0.29	–

(b)

Fig. 7.7 Character matrix and cladogram (a) to illustrate how phenetic distance measurements underestimate amount of change between pairs of taxa where there is homoplasy. The distances tabulated (b) show the absolute distance (as proportion of total number of characters compared) calculated both from cladistic branch length data and directly from a distance matrix. Pairwise comparisons in bold are underestimated by the distance method.

and then assigns character changes to each branch. Thus a character that acts as a synapomorphy for a deep branch but is reversed towards the tip counts as two changes between end members. The pairwise distance calculated by this method is greater than that derived from distance methods, as can be seen from the matrix in Fig. 7.7.

Distance matrix methods therefore underestimate the amount of morphological change that has occurred between end members in comparison to cladistic methods, and the more taxa and the more homoplasy, the greater will be the discrepancy between estimates based on distance and those based on parsimony. Foote's (1992a) analysis of blastozoan diversity is likely to have significantly underestimated morphological disparity for time intervals with large numbers of taxa (since homoplasy increases with the number of species included – Sanderson & Donoghue, 1989). Morphological change through time is best estimated by phylogenetic methods of pairwise comparison, not by phenetic methods.

Phylogenetics and morphological disparity

Cladogenic structure can often indicate whether morphological diversity is increasing through time. A simple hypothetical example helps to explain how this works (Fig. 7.8). A cladogram constructed independently of biostratigraphic data can be used to explore how morphological diversity proceeded through time. In the example given, the ages of the taxa have been substituted at the branch termini in place of names, with 1–4 indicating successive geological time intervals. Two contrasting cladogenic patterns are shown. In the pectinate tree, time intervals occur clustered and in sequence, as might be expected from a single evolving lineage. In this case the morphological divergence

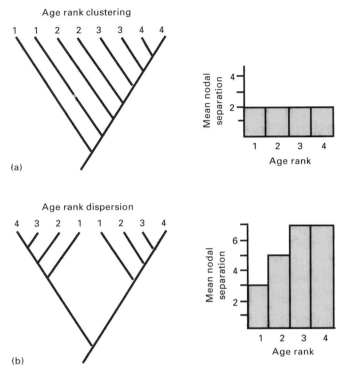

Fig. 7.8 Hypothetical example to demonstrate the use of mean nodal separation as an indicator of tree structure. The left-hand diagrams show cladograms in which 1, 2, 3, and 4 represent successive geological periods. In (a) taxa of the same age are adjacent whereas in (b) taxa of the same age become progressively more distant. Histograms to the right of each cladogram give the nodal separation between pairs of similar aged taxa.

between pairs of taxa of the same age is likely to be remain broadly similar over time, assuming that branch lengths are not wildly different. In contrast, the dichotomously branching cladogram has taxa of the same age progressively more dispersed over time, suggesting that morphological divergence is increasing over time. Thus the relative dispersal of taxa of the same age in a cladogram gives a crude measure of their morphological disparity.

This property can be quantified by calculating the mean number of nodes separating taxa of the same age. In the pectinate tree all pairs are separated by just two nodes, whereas in the dichotomous cladogram the number of nodes separating pairs increases from three to seven over the time interval. However, this approach is very susceptible to bias in sampling and differences in branch length (both terminal and internal) across the cladogram and only provides a very rough indicator of disparity. A more robust approach is to use the mean number of morphological changes calculated between all pairs of taxa of the same age as a measure of morphological diversity. This is basically the approach used by Foote (1992a), but in a phylogenetic context so that

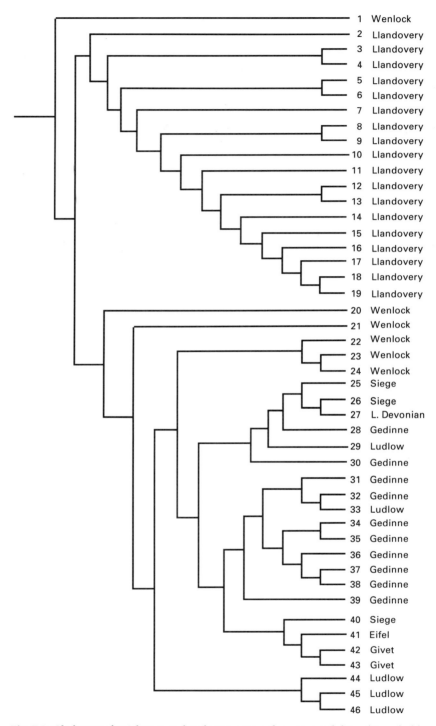

Fig. 7.9 Cladogram for Silurian and early Devonian phacopine trilobites (Ramskold & Werdelin, 1991). 80% majority rule has been used to select one tree out of the 32 equally parsimonious solutions found from the data presented by Ramskold & Werdelin. The geological age of each phenon is given. Note the excellent clustering of geological ages. Phena 1–46 are as follows: 1, *Podowrinella* (outgroup);

homoplasy can be taken into account. An analysis of trends in morphological disparity for phacopid trilobites provides an example.

Morphological disparity and phacopid trilobites

Ramskold & Werdelin (1991) provided a cladistic analysis for all well known Silurian–early Devonian trilobites of the subfamily Phacopinae based on 32 characters, mostly multistate continuous variables derived from morphometric analysis. Their resulting cladogram gave high resolution with few polychotomies, but identified 248 equally parsimonious arrangements for phena of the genus *Acernaspis*. My reanalysis of their data, after removing one highly unstable taxon in the cladogram, identified just three equally parsimonious topologies. Relationships among the other genera were more stable, but included one polychotomy, which was resolved using majority rule, whereby any branch supported by more than 80% of trees (26 of the 32 alternative topologies) was accepted. This method was used in the absence of any expert assessment of the relative reliability of characters in conflict. The resultant cladogram is shown in Fig. 7.9, with the geological ages of phena substituted for names.

Inspection of the cladogram structure shows that there is excellent clustering of geological age in the Llandovery and Wenlock, but that dispersal of taxa, with interleaving of phena of different geological age, becomes more common in the late Silurian. The relative dispersion can be estimated from the mean pairwise comparison of number of nodes separating taxa of the same age and standardizing for sample size (by dividing by $(n - 1)$, where n is the number of taxa compared). This indicates that on average the number of nodes separating phena of the same age increases during the Silurian, but drops in the early Devonian (Fig. 7.10).

This apparent pattern can be explored more rigorously by calculating the mean number of morphological changes that separate pairs of contemporary phena. The cladogram is used to assign character changes (including reversals and convergence) to each branch. Next the total branch length separating pairs of phena of the same age is derived by adding the character changes along the shortest path between pairs of taxa. The average is then calculated from the sum for every possible

---

(Continued.) 2, *Acernaspis woodburnensis*; 3, *Ac. quadrilineata*; 4, *Ac. konoverensis*; 5, *Ac. sufferta*; 6, *Ac. dispersa*; 7, *Ac. xynon*; 8, *Ac. rectifrons*; 9, *Ac. boltoni*; 10, *Ac. sororia*; 11, *Ac. incerta*; 12, *Ac. pulcher*; 13, *Ac. elliptifrons*; 14, *Ac. skidmorei*; 15, *Ac. salmoensis*; 16, *Ac. becsciensis*; 17, *Ac. superciliexcelsis*; 18, *Ac. mimica*; 19, *Ac. orestes*; 20, *Ananaspis* cf. *stokesii*; 21, *An. nuda*; 22, *An. amelangorum*; 23, *An.* sp. nov.; 24, *An. macdonaldi*; 25, *Kainops veles*; 26, *K. ekphymus*; 27, *K. microps*; 28, *K. invius*; 29, *K. guttulus*; 30, *K. raymondi*; 31, *Paciphacops crossleii*; 32, *P. serratus*; 33, *P. latigenalis*; 34, *P. eldredgei*; 35, *P.* sp. nov.; 36, *P. hudsonicus*; 37, *P. birdsongensis*; 38, *P. campbelli*; 39, *P. logani* 40, *Phacops claviger*; 41, *Viaphacops*; 42, *Phacops rana*; 43, *Phacops iowensis*; 44, *Ananaspis aspera*; 45, *An. decora*; 46, *An. fecunda*.

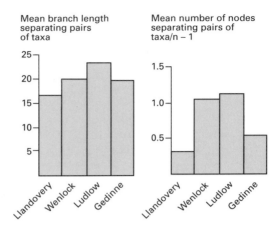

Fig. 7.10 Mean branch lengths (character changes) and mean number of nodes separating pairs of phacopine trilobite taxa of the same geological age (see text).

combination. When this is done for the phacopid cladogram, lowest morphological diversity is still found in the Llandovery sample, while highest diversity is seen in the Wenlock and Ludlow. There is a slight decrease in overall morphological diversity in Gedinnian phena, as before.

As with all measures of morphological diversity, sampling is all-important in determining whether we accept these results as valid. One possible source of error can arise from uneven density of sampling. Sanderson & Donoghue (1989) and Sanderson (1990) have observed that homoplasy increases with the number of taxa sampled, because denser sampling increases the chances of hidden homoplasy being revealed. In the phacopid example the Llandovery phena have the lowest mean morphological disparity, yet are also the most heavily sampled. Their low mean branch length (as measured between pairs of taxa) cannot therefore be an artefact arising from undetected homoplasy.

Another source of error arises if the phena included do not represent a random and unbiased selection of taxa. Ramskold & Werdelin (1991) used all adequately know Silurian phena, and the patterns derived for these are therefore likely to be genuine. However, of the Devonian taxa, only *Paciphacops* was treated in the same way (Devonian phacopids are much more numerous than Silurian ones), and there is the danger that the small drop in morphological diversity in the Gedinnian may simply reflect sampling bias. Only a more comprehensive survey of early Devonian phacopids will resolve this question.

Origination patterns

All taxa – be they monophyletic, paraphyletic, or polyphyletic – have an origination in time and space which corresponds to the first appearance of a phenon. Pearson (1992) referred to species that form part of an anagenic lineage as undergoing 'pseudospeciation', since their origin is the result of a taxonomist's arbitrary subdivision of a lineage. How great a problem this is remains to be seen, but it is likely that phena will form a complete spectrum from those based on subtle changes in ratios of morphological traits to those based on discrete and significant character acquisition.

The number of phena appearing in the fossil record has certainly fluctuated through time, but a clear discrimination between fluctuations caused by sampling of a variable fossil record, and those reflecting genuine biological phenomena, is not possible simply by counting taxa. In the previous example of cornute diversity in the early Palaeozoic (Fig. 7.3), the taxic approach would wrongly imply that cornutes underwent a major morphological radiation in the Arenig, whereas the phylogenetic approach shows that the observed pattern is much more likely to result from preservation failure in the late Cambrian and Tremadocian. The same preservational gap seems to be true of echinoderms in general (Smith, 1989), and Edgecombe (1992) has used similar reasoning to dampen the Ordovician radiation of trilobites. Thus, what at first sight appears to be a major burst of diversification in echinoderms, carpoids, and trilobites turns out to reflect, at least partially, sampling bias and ignorance of relationships (e.g. Edgecombe, 1992, p. 147).

It is also important to differentiate between immigration and origination, and again this comes from an awareness of phylogenetic relationships. The appearance of a taxon that does not seem to be closely related to any found earlier within that region potentially represents an immigration event. Archibold (1993) cites several examples of immigration events that can be recognized in the North American late Cretaceous and early Palaeocene mammal fauna.

The phylogenetic method allows sampling and preservation failure to be compensated by using sister group relationships to predict range extensions beyond that observed. Range extensions are invoked because of a mismatch between phylogeny and stratigraphy, when sister groups appear in the stratigraphic record at different times. If the morphological data are robust and both taxa are monophyletic, the origin of the younger taxon can be extended back in time to the first occurrence of the older taxon – as in the case of angiosperms discussed above (p. 140). Phylogenetic trees thus provide a closer estimate of the time of origination of a taxon where the fossil record is patchy, though they are not entirely unaffected by sampling inadequacies and may still underestimate time of origination.

Although the fossil record of post-Palaeozoic echinoids shows relatively good correspondence between cladogram branching order and stratigraphic appearance (Smith, 1981), there is one notable exception, the Echinothurioida. Echinothurioids are a group of flexible-tested echinoids today confined to the deep-sea environment. They have a very poor fossil record because of their imbricate skeleton and deep-water distribution. Thus there are only a handful of records of this group, the oldest being Upper Cretaceous. Yet morphological analysis shows them to be sister group to all other euechinoids (Jensen, 1981; Smith, 1981) and their ancestry must extend back at least to the first definitive record of euechinoids, which is late Triassic. This of course is only a latest possible divergence time for echinothurioids, but it does provide a more realistic estimate of the stratigraphic range of this group than simply taking the fossil record at face value.

Just how much of a difference does a phylogenetic approach make? In some cases there is an almost perfect match between the order in which taxa appear in the fossil record and their position on a cladogram. For example, Forey's (1991) recent analysis of 36 coelacanth genera ranging from Devonian to Recent, produced a cladogram in which the order of taxa was entirely congruent with their stratigraphic occurrence (Fig. 7.11; see also Cloutier, 1991). Here, even though the fossil record is one of the most notoriously patchy (with no record of a coelacanth between the Santonian and Recent), taxa are preserved in their correct order. Similarly, there seems to be excellent correspondence between the fossil record and the cladistic topology for groups such as the ornithomimosaurian (Barsbold & Osmolska, 1991) and ankylosaurian (Coombs & Maryanska, 1991) dinosaurs, and the fossil

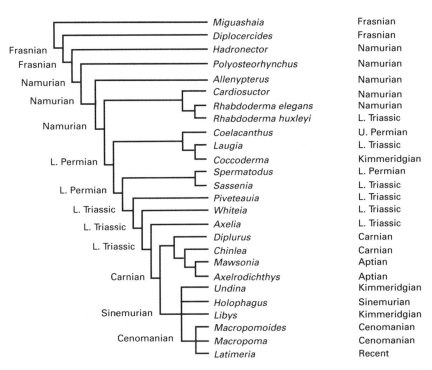

Fig. 7.11 Cladogram for actinistian fishes (coelacanths and their stem group) (Forey, 1991). Terminal taxa and their ages are given at branch tips.

record of equids (Evander, 1989). However, such excellent correspondence is not always to be expected. In eight (33%) of the 24 taxonomic groups looked at by Norell & Novacek (1992a), a taxon that was phylogenetically more derived appeared before the least derived taxon. Thus, although many taxa will appear in approximately the correct order stratigraphically, there are clearly going to be many exceptions.

Extinction patterns

Much has been published recently concerning extinction and its possible causes, but very little has been framed within a proper phylogenetic context. This is unfortunate because the last records of paraphyletic groups do not necessarily sample biological extinction, and phylogenetic hypotheses are needed to discriminate between disappearances that arise from nomenclatural convention and those that represent true biological extinction. Only monophyletic groups provide an unambiguous measure of lineage extinction.

One of the major problems associated with establishing extinction patterns is that of distinguishing local from global effects. In a shallow marine environment, for example, small-scale stratigraphic heterogeneity of lithofacies, each with its own distinctive fauna, is typical. Each change represents the local extinction of the earlier assemblage within the section, although taxa may simply have migrated to nearby areas along with the lithofacies. The large-scale sea-level fluctuations that cause major lithofacies reorganization over the continental shelf undoubtedly generate some total extinction (e.g. Westrop & Ludvigsen, 1987; Fortey, 1989; Hallam, 1989) but they also create significant problems of sampling. If we do not have lithofacies continuity within a region then taxa have clearly become extinct locally. However, how can we tell whether this event was enough to cause the total extinction of those taxa, or simply altered or reduced their overall geographic range? This question is particularly pertinent in the light of Valentine & Jablonski's (1991) findings that Pleistocene sea-level changes had little effect on the diversity of the marine fauna at any ecological level but generated extensive changes in phenon geographic ranges.

The early Silurian is a prime example of a significant gap in the fossil record of many groups. Of the eight families of cystoids known to pass from the Ordovician into the Silurian, only one is recorded in the Llandovery, and a further five appear in the Wenlock (Paul, 1982). Cocks (1988) noted similar absences of brachiopod taxa from the Llandovery, and Fortey (1989) listed a whole series of trilobite taxa that are present in the Ordovician and very latest Llandovery or Wenlock but are absent from the intervening period. Clearly there must have been refugia where these taxa survived, and the poor fossil record of the Llandovery is at least partially a result of sampling and preservation bias.

Provided there has been no significant character acquisition in any of these Lazarus lineages, there is little danger of mistaking a sampling gap for an extinction event. Problems arise if the gap represents a time interval long enough for morphological novelties to have accrued. When this occurs it offers taxonomists a convenient or uncontroversial breakpoint at which to separate a paraphyletic stem group from its derived sister group. Thus, for example, Milner (1990, p. 324) was able to claim that 'the boundary between the temnospondyls [a paraphyletic stem group] and their crown-group the lissamphibians, although arbitrary, is represented by a large gap in the fossil record and so presents little controversy'. In such cases there is a danger that nomenclatural changes will be more common at intervals where a period with a relatively good fossil record is followed by one with a relatively poor fossil record.

Paraphyly can be a major problem for the analysis of extinction patterns, as has been highlighted by Patterson & Smith (1987, 1988) and Archibold (1993). Previous analyses of a large data set of marine families and genera using the taxic method had revealed a number of major extinction peaks through the post-Palaeozoic (Raup & Sepkoski, 1984, 1986; Sepkoski & Raup, 1986). The distribution of these peaks suggested periodicity in extinction. However, when the echinoderm and fish taxa were examined critically, many of the peaks were found only in the non-monophyletic subset of data, not in the monophyletic subset, even though both subsets were approximately equal in size (Fig. 7.12). This suggested that extinction peaks were being exaggerated or even created by taxonomic practice.

Approximately 20% of disappearances at the major mass extinctions in echinoderms and fishes have been shown to be the result of paraphyly and taxonomic convention (Smith & Patterson, 1988). A similar estimate has been given for the number of trilobite families or subfamilies disappearing at the Cambro-Ordovician boundary (Fortey, 1989). In the latter case taxonomic specialization of trilobite workers may have exacerbated apparent extinctions across the Cambro-Ordovician boundary in that group (Briggs *et al.*, 1988; Edgecombe, 1992).

Paraphyletic and polyphyletic taxa should be broken into their monophyletic constituents to ensure that only monophyletic taxa are used and that any disappearances are true extinction events. The resolution afforded by this approach depends on the level of inclusiveness of the taxa. Maximum resolution is achieved if analysis is done at the level of phenon. At higher levels extinction patterns will become progressively less certain. If a clade, nominally assigned the rank of family, appears on either side of some gap in the record, this might be because every single lineage present below the boundary has its derived sister group above it, or because only one out of a large number of lineages passes across and is sister group to all the later taxa. In the former case the intervening period is simply a problem of sampling and preservation failure; in the latter case, the period marks a major

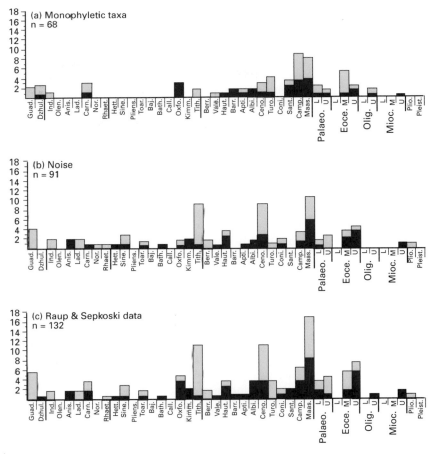

Fig. 7.12 Number of taxonomic extinctions in post-Palaeozoic fish (grey) and echinoderm (black) families. (a) The distribution of monophyletic clades (including monophyletic elements of paraphyletic groups and corrected last occurrences). (b) The distribution of last appearances of non-monophyletic taxa and monophyletic taxa that have been wrongly dated in the traditional database. (c) The results obtained using a traditional taxonomic database in which nonmonophyletic taxa abound. This formed a subset of the data used by Raup & Sepkoski (1986) to demonstrate periodicity of extinction. Peaks of extinction are apparent in the noise, but not in the clade data (Patterson & Smith, 1987).

and almost terminal drop in species diversity. Using higher taxa as a proxy for phenon-level patterns introduces uncertainty and may mask extinction peaks.

Archibold (1993, p. 25) distinguished between the disappearance of non-monophyletic taxa and that of metaspecies: 'metaspecies do not have artificially imposed limits but are constrained by apomorphies that they share with subsequent taxa and lack other apomorphies shared only by the subsequent taxa'. However, this is a classic definition of a paraphyletic group (see p. 76) and metaspecies are taxa whose boundaries are determined by default, according to the resolution afforded by available characters. When Archibold states that a metaspecies can 'evolve into another taxon via peripatry, bifurcation or anagenesis' what he is presumably claiming is that the origins of the descendant taxon are inferred to lie somewhere within the membership of the metaspecies. I can see no difference, other than scale, between this and the inference that birds (a monophyletic group) evolved from reptiles (a paraphyletic group). The disappearance of a metaspecies through inferred transformation is not extinction, and the 'evolutionary event' that this marks is the appearance of morphological novelty from which taxonomists can establish a further, more restricted set of individuals.

Taxonomic duration

Since only monophyletic taxa have both an origin and a biological extinction they are the only kind of group that can be said to have a duration other than by taxonomic convention. Traditional taxonomies, with their predominance of non-monophyletic groupings, are inappropriate for this sort of analysis. Since taxonomic rank is arbitrary, there is no justification for comparing the durations of taxa of equivalent rank within different clades, as this will reflect only the taxonomic practices prevalent in the different groups. Mean generic or familial durations need not carry biological significance, but may rather reflect how taxonomists have worked historically, the perceived importance of the group for stratigraphic correlation, and the relative morphological complexity of the different groups.

Ammonites, for example, as a result of their perceived stratigraphic utility, have tended to be highly subdivided by taxonomists, whereas less utilized groups such as gastropods are treated more conservatively. If resolution is important in defining a phenon's range, there should be some correlation between morphological complexity and longevity. Only Schopf *et al.* (1975) have ever tried to test this proposition. They found that there was a correlation between the perceived rate of evolution of a major taxonomic group and its morphological complexity, as measured by the number of terms applied to its skeleton. This implies that perceived rates of evolution may depend on the number of morphological characters available (or utilized) in the taxa studied.

Patterns of taxonomic longevity have been studied by constructing survivorship curves, a technique pioneered by Van Valen (1973a). Taxa within a more inclusive group are treated as if they formed a cohort originating at one point in time. The number of taxa surviving after specific time intervals is then plotted on semi-logarithmic axes. Van Valen (1973a) discovered that the great majority of his plots declined linearly and from this he argued that the probability of survival for taxa was independent of age – his 'Red Queen's hypothesis'. However, his data were based on traditional taxonomic compilations, and sample checks of echinoderm and fish taxa revealed that around 80% of the entries were non-monophyletic (Smith & Patterson, 1988). Furthermore, there is a crude correlation between the number of phena included and the duration of a taxon. If the great majority of taxa being used have their boundaries set arbitrarily, the suspicion arises that taxonomic duration is to a large extent controlled by taxonomic practices (Arnold, 1982; Smith & Patterson, 1988). Survivorship plots appear linear because of random factors in the allocation of species to higher taxa: the more speciose a group, the more likely it will be split by a taxonomist. This kind of random process can generate a linear correlation in a taxonomic survivorship plot (Anderson, 1974; Anderson & Anderson, 1975).

Raup (1978) subsequently developed and refined the application of survivorship curves to palaeontological data by plotting the fate of cohorts of taxa originating during successive geological periods through time. This approach has subsequently been applied by Hoffman & Kitchell (1984), Jones & Nicol (1986), Raup (1987a), and Foote (1988). Foote, for example, used cohort analysis of trilobite generic durations to compare those originating in the Cambrian and Ordovician. He found that Ordovician taxa survived on average three times longer than Cambrian taxa (Fig. 7.13). Among the possible causes that might have given rise to this pattern, Foote considered taxonomic artefact. Since trilobites are much more important biostratigraphic tools in the Cambrian than in the Ordovician, they have possibly been more finely subdivided by taxonomists.

For example, late Cambrian trilobites have been split into a succession of faunas, each of which shows a broadly similar range of morphologies (Palmer, 1965). These faunas define biomeres and each was

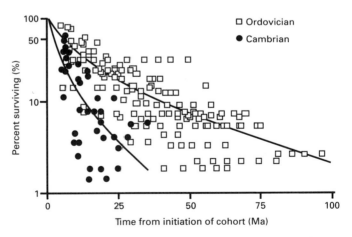

Fig. 7.13 Survivorship curves calculated for Cambrian and Ordovician cohorts of trilobite (Foote, 1988). Each point represents a cohort and the percentage surviving is plotted on a logarithmic scale. The best fit curves for Cambrian (lower curve) and Ordovician (upper curve) are shown.

assumed to represent an independent clade that became extinct at the end of its biomere, with only one or two generalist taxa surviving through to the next period. Similar morphologies arising in successive biomeres were seen as convergent and the result of iterative evolution from some common, slowly evolving, generalized ancestor rather than continuation of the same lineage. However, Fortey (1983) suggested that the similarity seen between faunas before and after an extinction event reflected true homology, not convergence. Under his model, taxa survived in more marginal (unsampled) positions and recolonized the shelf after each local extinction event. Only detailed cladistic analysis will resolve whether successive homeomorphs are in fact sister groups or the result of iterative convergence. However, the suspicion is that biostratigraphy is being used to oversplit trilobite ranges in the Cambrian as compared to the Ordovician.

Survivorship analysis has been refined further by employing a simple correction factor to take account of variation in real-time extinction rates (Pearson, 1992). Survivorship curves require that the group under consideration is homogeneous both in terms of composition and probability of extinction. Taxa arising at different periods in the geological past have experienced different extinction probabilities, therefore they cannot be considered homogeneous and are not appropriate for survivorship analysis without correction (Levinton, 1988, p. 428). Pearson developed a correction factor for this and was able to demonstrate that the probability of extinction for Palaeogene planktonic foraminifers increases with longevity: older phena are more likely to go extinct at events than newer taxa (Fig. 7.14). Does this indicate that phena are like individuals and have a predetermined senescence? Pearson thought not and attributed this pattern to the way in which taxonomists have chopped up chronoclines into a series or overlapping named segments. Indeed he seriously questioned whether any survivorship analysis to date reflects anything other than taxonomic artefact.

Monophyletic taxa have meaningful taxonomic durations, yet the arbitrariness of taxonomic rank makes comparison across taxa or over time difficult to interpret. Comparison of sister taxa seems the only valid way to compare clade duration. The most informative approach would be to compare pairs of sister groups from a cladistic analysis that each differ in the same attribute (see Jensen, 1990). Sister groups of late Cretaceous molluscs in which one clade has a planktotrophic larval stage and the other a lecithotrophic larval stage, for example, could be compared to discover whether planktotrophic clades have a significantly longer taxonomic duration on average than their lecithotrophic sister groups. As far as I am aware, no such comparisons have yet been made using palaeontological data, although a similar approach (comparing relative species richness of pairs of sister taxa that differ in one attribute) is applied to neontological data (e.g. Nee *et al.*, 1992 for birds).

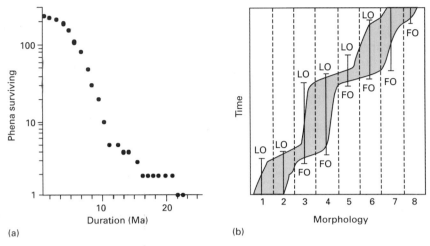

Fig. 7.14 (a) Survivorship curve for Palaeogene planktonic foraminiferans calculated using Pearson's corrected survivorship score. Note the sigmoidal shape of this curve. (b) Arbitrary taxonomic subdivision of lineages can lead to convex survivorship curves. The stippled zone marks the intrapopulation variation seen in an evolving lineage. Vertical dashes represent the arbitrary boundaries set by taxonomists wishing to use phena for biostratigraphic purposes; nominal phena are numbered 1–8. LO, last occurrence datum, FO, first occurrence datum. For coexisting species the older phenon tends to disappear (through pseudoextinction) from the fossil record before the younger. (Both diagrams from Pearson, 1992).

Species as phena: stasis and phyletic evolution

The widely held conflation of taxonomists' phena with biological species and thus 'individuals' has led to species durations being perceived as having particular importance. However, since phena in the fossil record belong to one of three distinct kinds of groups (single populations, monophyletic taxa, and metataxa) they cannot be considered commensurate entities. Furthermore, the term 'species' is simply a tag of taxonomic convention for a level of inclusiveness that is determined by the systematist's ability (or preferences) and the resolution allowed by available morphological data. The perceived durations of phena will therefore be influenced by several factors: phena based on single populations are likely to be inadequately sampled, while metataxa may have their boundaries established by taxonomic convention or by chance gaps in the fossil record. Thus phena are heterogeneous entities and analysis of their durations may be largely meaningless.

Phena do of course provide information about the history of character acquisition and transformation through time. By combining cladistic and stratigraphic data it is also possible to distinguish between patterns of morphological anagenesis and cladogenesis (Archibold, 1993).

Much of the argument over punctuated versus gradual evolution has hinged on the analysis of individual characters. However, single character analyses may be misleading. Cheetham (1987) provided one of the few analyses of morphological evolution based on an extensive suite of characters (46 characters in nine species). This study found strong evidence for stasis within phena; only a few temporal trends in characters could be identified and these were unrelated to inferred ancestor–descendant transformations. Stanley & Yang (1987) compared 19 early Pliocene bivalves with their nearest living relatives based on another massive compilation of morphometric data. For each pair, 24 variables were compared, as were eigenshape values for shell outline. They found that the variation between populations aged 4 Ma and extant populations was of the same order of magnitude as that separating 'conspecific' Recent populations from different geographic localities. They therefore concluded that evolution (of the characters measured) had followed a weakly zigzag course, yielding only trivial net trends.

These two studies together strongly suggest that character stasis is statistically more predominant in the fossil record than phyletic gradualism. Other workers, however, provide amply documented evidence that phena can be differentiated only on arbitrary division of morphological continua (Baarli, 1986; Rose & Bown, 1986; Sheldon, 1987; Tabachnick & Bookstein, 1990; Pearson, 1992). Rose & Bown (1986, p. 127), for example, claimed that, in the groups with which they were familiar, 'a dense fossil record often reveals gradual transformations between species both within lineages and at branching events thus complicating species recognition. In the best documented lineages, this compels us either to impose arbitrary species boundaries or none at all'.

The deep-sea record of planktonic foraminiferans and radiolarians probably represents the best source of continuous sequences and abundant material from which to examine rates of morphological change. Lazarus (1986) provided a detailed morphometric analysis of the oceanic radiolarian *Pterocanium* from deep-sea cores which appears to record a cladogenic event. Lazarus ran a discriminant analysis sample by sample, based on 32 measured variables. This revealed that, within a space of approximately 50 000 years, there was a change from a single polymodal population to two discrete, non-overlapping populations. Thereafter relatively rapid phyletic evolution continued for about the next 600 000 years. During the following 1.8 Ma there seemed to be much less net morphological change in the lineage, although no detailed sampling was carried out and significant variation on a small to medium timescale cannot be ruled out. Lazarus pointed out that the pattern of morphological evolution in *Pterocanium* combined both gradualistic and punctuational aspects.

Malmgren *et al.* (1983) provided one of the earliest really well documented examples of step-like evolutionary change. They used both overall size and a multivariate measure of overall shape to examine morphological evolution in a lineage of the foraminifer *Globorotalia* over the last 10 Ma, with data taken from a single deep-sea core. This appeared to show a relatively sharp transition at around 5.2 Ma, separating two discrete morphotypes, *G. plesiotumida* and *G. tumida* (Fig. 7.15a).

However, their data were reanalysed by Bookstein (1987) to test whether the pattern could be distinguished from that of a symmetrical random walk. He used two approaches, a scaled-maximum test that is independent of timescale, and an evolutionary rate approach that is dependent on timescale. Bookstein concluded that the *Globorotalia* data were consistent with the lineage having undergone a random walk, showing neither punctuation nor anagenesis. This of course does not prove that evolution in *Globorotalia* was the result of random drift, only that the factors involved are many and complex.

MacLeod's (1991) re-evaluation of the same data is even more

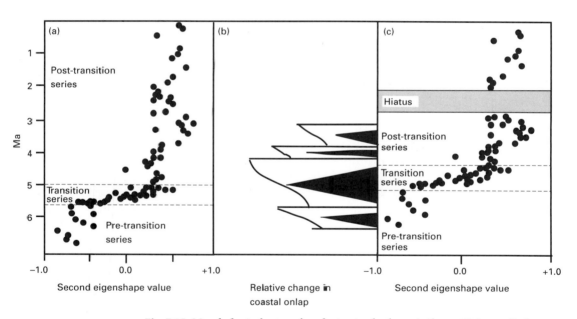

Fig. 7.15 Morphological rates of evolution in the foraminiferan *Globorotalia* from DSDP site 214 (redrawn from MacLeod, 1991). (a) The data on shape (based on multivariate morphometric analysis) as presented by Malmgren *et al.* (1983), uncorrected for variation in sedimentation rates. A sharply defined transitional series is identified. (b) Coastal onlap curve showing the correspondence between a major onlap sequence and the transitional period. The black wedges indicate the temporal duration and relative magnitude of stratigraphic condensation events expected: the larger the wedge, the greater the expected condensation. Maximum condensation occurs at the inflection point on the coastal onlap curve. (c) Revised shape variation in *Globorotalia* after correction for stratigraphic condensation by graphic correlation. The transitional period becomes much more gradational and there is no statistically significant change in rates within the period shown below the hiatus.

crippling to the original hypothesis of punctuated anagenesis. He used graphic correlation to identify hiatuses and correct for variation in rates of sedimentation in the succession. This identified several discontinuities at which substantial changes in the relative rate of sediment accumulation had taken place. In particular, he found that the key transitional phase between *G. plesiotumida* and *G. tumida* that defines the punctuated anagenesis event lies entirely within an interval of relative sedimentary condensation (Fig. 7.15b).

MacLeod constructed, from his graphic correlation, a revised timescale estimate for the *Globorotalia* lineage, and recalculated the rates of evolution. Using Bookstein's test, he found evidence for a slight increase in the speed of morphological change in the transitional interval and for significant differences in the rate of morphological change (as measured by the signed reduced-speed metric) between four successive series. However, he pointed out that these four series were measured at different levels of temporal resolution and that there was a correlation between magnitude of morphological rates of change and temporal resolution (also previously documented by Gingerich, 1983, and Gould, 1984).

A bootstrap test, similar to that used by Kitchell *et al.* (1987), showed that the observed rate of size and shape change for the transition series is not significantly different from the evolutionary rates that characterize other parts of the lineage. MacLeod concluded that the morphological transition between these two species took place along a gradual continuum (Fig. 7.15c). He further pointed out that other planktonic foraminifer lineages that are claimed to show punctuated evolution, do so at or close to major palaeo-oceanographic events that have affected sedimentation rates in the deep sea.

As Fortey (1985, 1988) has demonstrated, much of the debate about phyletic gradualism versus punctuated equilibrium depends on the quality of the fossil record and the density of sampling. Despite much research effort, we seem no closer to deciding which mode predominates. Instead the whole question has become overwhelmed by problems of sampling and insufficient resolution, both spatial and temporal, provided by the fossil record. Detailed studies of sections can give high temporal resolution through fine-scale stratigraphic collection, but lack sufficient spatial resolution to distinguish local immigration and emigration events from speciation. On the other hand, studies extending over a wide geographic area suffer from our general inability to correlate on a sufficiently fine time-scale between sections.

Although there is evidence that both rapid and gradual modes of character evolution occur, each can be interpreted differently. Thus gradual change to the punctuationist is not regarded as speciation, while punctuated change to the gradualist is confirmation of the inadequacies of the fossil record. Sampling problems caused by variation in rates of sediment accumulation are particularly acute and are rarely accounted for adequately (MacLeod, 1991; MacLeod & Keller, 1991). Finally, since many phena in the fossil record are certain to be defined solely on plesiomorphic traits, much of the discussion of species durations and speciation rates is additionally plagued by problems of paraphyly and arbitrary boundaries.

Rates of evolution

There are three kinds of rates of evolution that can be calculated: rates of morphological evolution, rates of taxic evolution, and rates of genomic evolution (Schoch, 1986; Raup, 1987b). All are best approached from the basis of phylogenetic trees.

Rates of morphological evolution

Rates of morphological evolution through geological time were first examined seriously in the pioneering work of Simpson (1944, 1949). The most direct approach is to measure the amount of morphological change that has taken place along branches of a phylogenetic tree. In the past, geological age has simply been used to order taxa and thus establish rates of morphological change without constructing a secure phylogeny (e.g. Westoll, 1949; Schaeffer, 1952). Consequently these studies offer only a crude measure of morphological change. However, recently Cloutier (1991) has attempted a more comprehensive study of coelacanth evolution that begins from a phylogenetic analysis.

Cloutier constructed a cladogram for 31 actinistian fishes, which ranged in age from late Devonian to Recent and included most of the better known phena of that clade. The analysis was based on 75 skeletal characteristics and generated two equally parsimonious clado-grams that differed in only one small detail (Fig. 7.16a). Cloutier then used the stratigraphic distribution of the 31 taxa and the results of his cladistic analysis to construct a phylogenetic tree (Fig. 7.16b). From the cladistic analysis he was able to assign character changes to each branch of the tree. There were 101 character-state changes along a phylogenetic pathway composed of 17 cladogenetic events leading from the basal node to the present-day *Latimeria*. Assuming that the tree is a true reflection of the clade's history and that the character distributions are a true reflection of character changes (i.e. an unbiased sample), Cloutier was then able to calculate rates of morphological change through geological time for this clade.

Three different metrics were used: (i) the number of changes between consecutive cladogenic events; (ii) the number of changes along the stem lineage over consecutive fixed periods of time (50 Ma); and (iii) the number of changes during consecutive geological periods. For all three metrics, the total number of changes assigned to the stem line-age pathway (i.e. the direct line of descent leading from the basal node

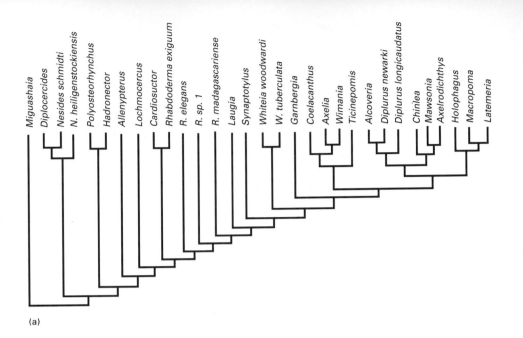

(a)

Devonian		Carboniferous		Permian		Triassic		Jurassic			Cretaceous		Tertiary			
M	U	L	M	L	U	M	U	L	M	U	L	U	P	E	O	M

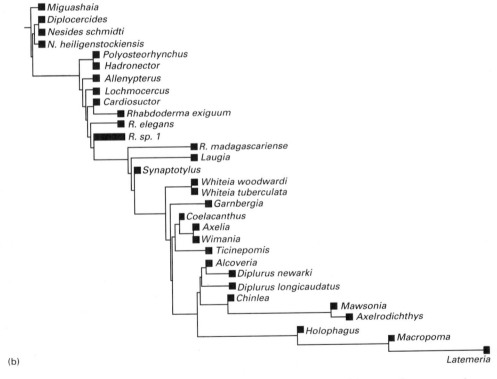

(b)

Fig. 7.16 Actinistian phylogeny. (a) Cladogram derived from 75 characters and one of two equally parsimonious solutions found. The alternative topology treats *Polyosteorhynchus*, *Hadronector* and the clade beginning with *Allenypterus* as a trichotomy. (b) Phylogenetic tree constructed from observed biostratigraphic ranges and the cladogram branching pattern. (Both redrawn from Cloutier, 1991).

on the cladogram to the extant *Latimeria*) was divided by the duration of the period itself.

Cloutier found that all three approaches gave him approximately the same result, namely a relatively high rate of evolution early on in the clade's history that decreased exponentially through time (Fig. 7.17). Note that, in contrast, the morphological disparity of the group actually increases up to the mid-Cretaceous, as measured using the phylogenetic method of mean pair-wise comparison outlined above. Observed phenon diversity also increases to a maximum in the Triassic and thereafter declines to the present day. Thus rates of morphological diversification are clearly not correlated with rates of phenon diversity in this example.

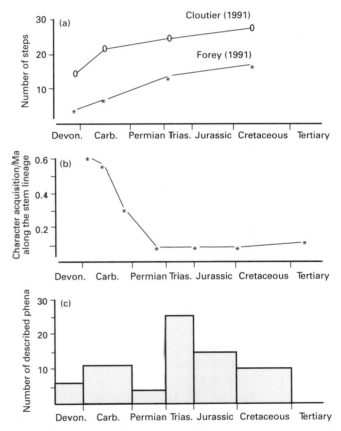

Fig. 7.17 Rates of evolution and morphological disparity for actinistian fishes. (a) Mean pair-wise cladistic distance between taxa calculated as described on p. 173. Two cladistic analyses were used, that of Forey (1991), based on 56 characters, and that of Cloutier (1991), based on 75 characters. Both show the same general increase in morphological disparity up to the mid-Cretaceous. The four points were calculated using: (i) late Devonian taxa, (ii) Namurian taxa, (iii) Lower and Middle Triassic taxa, and (iv) Aptian–Cenomanian taxa. (b) Rate of morphological change, calculated as the number of character changes along the stem group lineage divided by the length of the branch in Ma (Cloutier, 1991). (c) Numbers of described phena of actinistians through geological time (Forey, 1991).

Rates of genomic evolution

Smith *et al.* (1992) used basically the same approach as Cloutier to document and compare rates of genomic and morphologic evolution through geological time for lineages of post-Palaeozoic echinoids. Ten extant taxa were selected and their large subunit ribosomal RNA extracted and sequenced. These partial sequences were then aligned and used to construct a molecular phylogenetic tree using parsimony. The same taxa were also scored for a total of 81 skeletal characters and these then used to construct a morphological phylogeny, also based on parsimony. Since both the morphological and molecular phylogenies were identical in topology, the results could be combined and a single tree constructed and calibrated against the fossil record (Fig. 7.18).

This tree provided 18 independent branches for which the numbers of morphological and molecular changes could be inferred. The duration of each branch was estimated directly from the fossil record, giving three variables that could be compared in each branch. Since none of the variables studied is normally distributed, the most appropriate procedure was to convert the data into rankings. Smith *et al.* used the non-parametric rank–order correlation, or Spearman correlation, to compare rates of morphological evolution and rates of molecular evolution against duration of branch, and rates of molecular evolution against morphological evolution. Both morphological and molecular evolution against time showed a similar level of correspondence, suggesting that rates of ribosomal RNA evolution are no more clock-like than morphological evolution, at least when averaged over these 10 taxa (Fig. 7.19).

Rates of taxic evolution

Since taxonomic rank is arbitrary, the only meaningful approach to taxic diversity is through comparison of phenon diversity between sister groups (Cracraft, 1984; Novacek & Norell, 1982). Sister groups represent clades that have arisen from a common ancestor and thus have the same age of origin. Differences in branching rates between pairs of sister taxa can then be tested using standard statistical techniques. This is sometimes done by simply comparing the total number of phena existing in each clade after a certain length of time. However, observing that two sister taxa differ significantly in the numbers of included phena does not prove that those differences arose at the dichotomy in question. For example, rates of diversification might be similar in the early history of both sister groups and change markedly in one clade only some time after its start. In that case the change in branching rate would not have been identified to the correct level in the hierarchy.

With a phylogenetic tree it is immediately apparent where branching rates are high (many short internal branches at the base of a clade) and

(a) Molecular cladogram

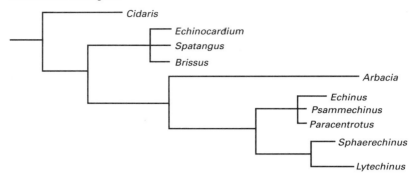

(b) Morphological cladogram

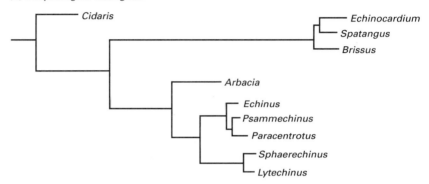

(c) Phylogenetic tree

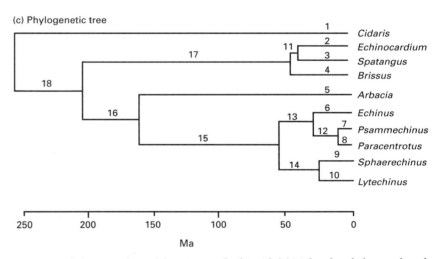

Fig. 7.18 Cladograms derived for 10 taxa of echinoid. (a) Molecular cladogram based on the first 380 bases from the 5′ end of the large subunit ribosomal RNA molecule (53 variable sites). (b) Morphological cladogram based on 81 morphological characters. The two cladograms have the same topology but different branch lengths. (c) Calibrated phylogenetic tree constructed using the first appearance of fossils that can be assigned to one or other sister group. For each of the 18 branches numbered in the cladogram it is possible to estimate: (i) the duration of the branch (in Ma), (ii) the number of molecular changes in the rRNA sequence that have accrued, and (iii) the number of morphological changes that have accrued (Smith *et al.*, 1992).

where they are low (long widely-spaced internal branches). It is then relatively easy to calculate the mean branch-length in the basal part of sister groups to see whether there is a significant difference. For example, Vrba (1980, 1984) documented evolutionary rates in African bovids and showed that the sister groups Alcelaphini and Aepycerotini contained significantly different numbers of phena after the same length of time. But was this difference initiated at their time of divergence or at some later date? Sanderson & Bharathan (1993) used a phylogenetic tree to show that branching rates at the base of the two sister taxa differed by a factor of about four, implying that differences in taxic diversification were indeed initiated at that level in the hier-

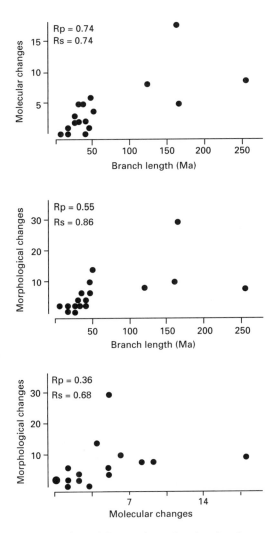

Fig. 7.19 Bivariate scatter plots of the numbers of molecular changes, numbers of morphological changes and branch length duration (in Ma) for the 18 branches in the echinoid tree identified in Fig. 6.17. Rp, Pearson correlation coefficient; Rs, Spearman rank correlation coefficient (Smith *et al.*, 1992).

archy. However, a maximum likelihood approach did not show that this difference was significant at the 0.05 level, and it could have arisen by chance alone.

Neither simple diversity patterns nor cladograms on their own provide as accurate an assessment of rates of branching as is obtained when both types of data are combined into a phylogenetic tree (Sanderson & Bharathan, 1993). Whether such differences have biological significance is another question and must be argued case by case. Higher taxic rates of diversification may simply arise where one sister group has acquired better preservation potential (e.g. a more robust skeleton, or a shelf as opposed to deep-sea habitat), or when a complex apomorphy has arisen that allows finer taxonomic subdivision.

Fossils and biogeographic patterns

Until recently, historical biogeography has largely involved the construction of scenarios. The palaeogeographic location of the oldest known fossil in a taxon was commonly deemed to be the site of origin for that taxon. A direct reading of the fossil record was then used to map the spread of the clade, through dispersion, from its centre of origin.

The concept of 'centres of origin' fell into discredit as a result of the extension of cladistic methodology into the field of biogeography and the emphasis this placed on the search for general biogeographic patterns common to many clades (e.g. Nelsen, 1978a; Nelsen & Platnick, 1981; Humphries & Parenti, 1986). However, all clades have to originate somewhere and cladistic methods can now offer a rigorous approach to understanding their individual biogeographic history.

Cladograms that include all known members of a clade, fossil as well as Recent, provide our best hope for documenting changes in distribution and establishing the timing of these events. This is because the addition of fossils increases the sampling density of areas of distribution occupied by a clade, increasing the chances that any convergence (through dispersal) will be recognized. Another advantage is that extinct taxa will often plot as branches more basal in the cladogram than any extant taxon, and thus provide sampled distributions closer to the origin of the clade. Mooi (1990) provided a cladogram for clypeasteroid echinoids (Fig. 7.20). The Clypeasterina, Laganina, and Scutellina each have a more or less cosmopolitan distribution today and their living sister taxon, the Cassiduloida, though restricted to only a few areas of the globe now, had an equally cosmopolitan distribution in the past. Since the Clypeasterina, Laganina, and Scutellina are a monophyletic group, they presumably had to have an origin somewhere. Can we specify where this may have been? Mooi (1990) and others identified only two extinct plesions to the Clypeasteroida clade: (i) oligopygids, comprising just two genera and a large number

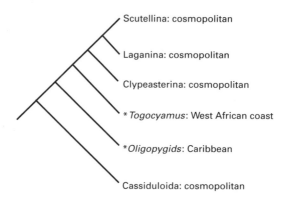

Fig. 7.20 Cladistic relationships for clypeasteroid echinoids with their observed biogeographic equatorial distribution (Mooi, 1990). Two extinct taxa (asterisked*) are sister group to crown-group clypeasteroids and allow the ancestral biogeographic distribution for the crown group to be more narrowly defined. The other four taxa are extant.

of nominal species; and (ii) *Togocyamus*, comprising two phena. All oligopygids are restricted to the Caribbean and are Eocene in age, while *Togocyamus* is restricted to the north-western African coastal belt and is Palaeocene in age. From cladogram topology, irrespective of any age considerations, evidence favours a Caribbean/West African origin for clypeasteroids. Coincidentally, these two areas lay much closer together in the early Palaeocene and have since drifted apart with the opening of the Atlantic. Adding fossils to a biogeographic analysis may therefore point more precisely to a clade's region of origin.

A more rigorous cladistic approach for identifying ancestral areas has been developed by Bremer (1992):

1 First a cladogram is constructed for a clade on the basis of morphological features.

2 Next the geographic regions that each member of the clade occurs in are identified.

3 Each area is then treated as a binary character and optimized on the cladogram. These areas must either have been part of the original ancestral area in which the clade arose, or represent areas into which the clade has subsequently dispersed. Each alternative is examined in turn:

(a) First, each area is treated as if it were part of the original ancestral area and the minimum number of losses (i.e. regional extinctions) that are implied from the cladogram structure (L) are calculated.

(b) Next, each area is treated as if it were an area that has only subsequently been inhabited, through immigration, and the minimum number of such appearances (gains) implied by the cladogram structure (G) is also calculated.

4 The gain–loss quotient (G/L) then gives a measure of the relative probability of that area being part of the ancestral range. Bremer's method is equally applicable to entirely fossil examples.

The geographic distribution of phacopid trilobites through time has been examined by Ramskold & Werdelin (1991). They constructed a cladogram for Silurian and Devonian taxa based on morphological evidence and then replaced the taxonomic names at the termini of the cladogram by the regions where the phena came from (Fig. 7.21). This demonstrates that taxa at the basal nodes are from areas concentrated around the Iapetus Ocean; there are no representatives from Gondwana. By node B, phacopids are widely distributed, while from node C upwards almost all taxa are from eastern Laurentia (i.e. USA). The exceptions are two small monophyletic groups confined to Australia and a single Czechoslovakian species. Ramskold & Werdelin concluded that all post-Ludlow phacopines originated in Laurentia and subsequently expanded to other areas.

If Bremer's (1992) method of determining ancestral areas is applied to Ramskold & Werdelin's cladogram, the gain–loss quotient (G/L) can be calculated for each of the 10 geographic areas represented (Table 7.1). In order to demonstrate the approach I have selected just one of their most parsimonious solutions. Alternative topologies will certainly give slightly different results, but should not make any major difference to the conclusions. Gain–loss quotients have been calculated for three nodes (labelled A, B, and C in Fig. 7.21). The most probable ancestral areas for the basal node A are Greenland and Scotland. Noth were part of the same terrain at that time and have known Llandovery phena. The ancestral area for node B includes Ireland and, less probably, Wales, while that for node C is clearly identified as the USA, as Ramskold & Werdelin (1991) predicted.

The use of fossils for identifying ancestral areas has been heavily and justifiably criticized for simply assuming that the oldest member of a group is ancestral and thus represents its origin. However, fossil data are important for biogeographic studies when analysed using appropriate cladistic techniques. The cladistic methods now available allow a rigorous approach to the recognition of ancestral areas and the history of clade dispersal.

Summary

A major goal of palaeontology is the recognition of general patterns of evolution through time. These then provide the basic evidence from which processes of evolution are deduced. If we are to refine our understanding of evolutionary processes, we must strive to construct evolutionary patterns using data that are as free of artefact as possible. In palaeontology artefact arises from three sources: the use of non-monophyletic taxa, the assumption that categorical rank is commensurate across taxa, and the fluctuating quality of the fossil record.

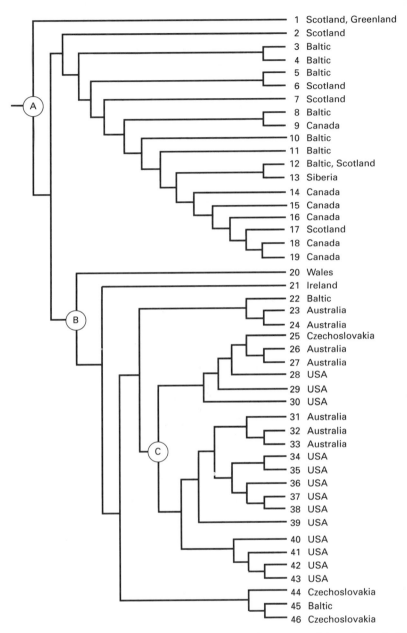

Fig. 7.21 Cladogram for Silurian to early Devonian phacopine trilobites (Ramskold & Werdelin, 1991). Taxa 1–46 are given in the caption to Fig. 7.9. The geographic occurrence of each terminal taxon is listed. (See Fig. 7.9 and text for further details.)

There can be little doubt that cladistic methodology offers the best tool for documenting meaningful biological patterns in the fossil record. Cladistics minimizes artefact by recognizing only monophyletic taxa. Non-monophyletic taxa, as arbitrary constructs, need to be eradicated from our database because the patterns that they reveal have

Table 7.1 Gain–Loss (G/L) quotients calculated for three nodes on one of the equally parsimonious cladograms derived for phacopid trilobites based on the data presented by Ramskold & Werdelin (1991) (see also p. 173). Nodes A, B, and C are indicated on Fig. 7.21. For each of the 10 geographic areas, the number of gains (G) implied by optimizing the observed distribution onto the cladogram, assuming it was not part of the ancestral range, and the number of losses (L) implied, assuming that it was part of the ancestral range, are calculated. The ratio G/L then gives a relative measure of how likely the area is to have been part of the original ancestral area. AA = the G/L quotient rescaled to a maximum value of 1 by dividing by the largest G/L value

Area	Node A			Node B			Node C			
	L	G	G/L	L	G	G/L	L	G	G/L	AA
Scotland	11	6	0.55	–	–	–	–	–	–	–
Canada	10	2	0.20	–	–	–	–	–	–	–
Greenland	1	1	1.00	–	–	–	–	–	–	–
Siberia	11	1	0.09	–	–	–	–	–	–	–
Wales	3	1	0.33	1	1	1.00	–	–	–	–
Ireland	4	1	0.25	2	1	0.50	–	–	–	–
Baltic	13	4	0.31	6	2	0.33	–	–	–	–
Australia	13	3	0.23	11	3	0.27	7	2	0.29	0.14
Czechoslovakia	10	2	0.20	9	2	0.22	5	1	0.20	0.10
USA	8	1	0.12	6	1	0.16	1	2	2.00	1.00

more to do with the way in which taxonomists have worked than about how evolution has proceeded. Taxonomic rank is used solely to indicate relative inclusiveness and the only truly comparable clades are sister taxa. Finally, combining cladistic analysis and biostratigraphic data into trees provides the best method to compensate for the sampling inadequacies of the fossil record. Thus phylogenetic trees provide our best evidence for reconstructing the individual histories of clades, and, by comparing tree structure across a number of taxa, general patterns of evolution can be discovered.

The potential applications of well-constructed phylogenetic trees are numerous, as I have tried to show in this chapter. Others will surely be found. My main hope is that this book will encourage the construction of more phylogenetic trees in the search for a better understanding of evolutionary patterns and processes.

References

Adams, E.N. (1972) Consensus techniques and the comparison of taxonomic trees. *Systematic Zoology*, **21**, 390–397.

Adams, E.N. (1986) N-trees as nestings: complexity, similarity, and consensus. *Journal of Classification*, **3**, 299–317.

Allard, M.W. & Miyamoto, M.M. (1992) Perspective: testing phylogenetic approaches with empirical data, as illustrated with the parsimony method. *Molecular Biology and Evolution*, **9**, 778–786.

Allison, P.A. (1988) The role of anoxia in the decay and mineralization of proteinaceous macrofossils. *Paleobiology*, **14**, 139–154.

Allison, P.A. & Briggs, D.E.G. (1993) Paleolatitudinal sampling bias, Phanerozoic species diversity, and the end-Permian extinction. *Geology*, **93**, 65–68.

Allmon, W.D. (1989) Palaeontological completeness of the record of lower Tertiary mollusks, U.S. Gulf and Atlantic Coastal Plains: implications for phylogenetic studies. *Historical Biology*, **3**, 141–158.

Anders, M.H., Kruger, S.W. & Sadler, P.M. (1987) A new look at sedimentation rates and the completeness of the stratigraphic record. *Journal of Geology*, **95**, 1–14.

Anderson, S. (1974) Patterns of faunal evolution. *Annual Reviews in Biology*, **49**, 311–332.

Anderson, S. & Anderson, C.S. (1975) Three Monte Carlo models of faunal evolution. *American Museum Novitates*, **2563**, 1–6.

Archibold, J.D. (1993) The importance of phylogenetic analysis for the assessment of species turnover: a case history of Paleocene mammals in North America. *Paleobiology*, **19**, 1–27.

Archibold, J.D. & Bryant, L.J. (1990) Differential Cretaceous-Tertiary extinctions of non-marine vertebrates: evidence from northeastern Montana. In: V.L. Sharpton & P.D. Ward (eds) *Global Catastrophes in Earth History: an Interdisciplinary Conference in Impacts, Volcanism, and Mass Mortality.* Special Paper, Geological Society of America, **247**, 549–562.

Archie, J.W. (1985) Methods for coding variable morphological features for numerical taxonomic analysis. *Systematic Zoology*, **34**, 326–345.

Arkell, W.J. (1957) Introduction to Mesozoic Ammonoidea. In: R.C. Moores (ed.) *Treatise on Invertebrate Paleontology. Part L. Mollusca 4, Cephalopoda, Ammonoidea*, pp. L81–128. Geological Society Of America and University of Kansas Press, Lawrence.

Arnold, A.J. (1982) Species survivorship in the Cenozoic Globigerinida. *Proceedings of the Third North American Paleontological Convention*, **2**, 9–12.

Avise, J.C. (1989) Gene trees and organismal histories: a phylogenetic approach to population biology. *Evolution*, **43**, 1192–1208.

Avise, J.C. & Ball, R.M. (1990) Principles of genealogical concordance in species concepts and biological taxonomy. *Oxford Surveys in Evolutionary Biology*, **7**, 45–67.

Avise, J.C., Arnold, J., Ball, R.M., Bermingham, E., Lamb, T., Neigel, J.E., Roeb, C. & Saunders, N.C. (1987) Intraspecific phylogeography: the mitochondrial

DNA bridge between population genetics and systematics. *Annual Reviews in Ecology and Systematics*, **18**, 489–522.

Ax, P. (1987) *The Phylogenetic System*. Wiley & Sons, Chichester.

Baarli, B.G. (1986) A biometric re-evaluation of the Silurian brachiopod lineage. *Stricklandia lens/S. laevis*. *Palaeontology*, **29**, 187–205.

Baker, A.N., Rowe, F.W.E. & Clark, H.E.S. (1986) A new class of Echinodermata from New Caledonia. *Nature*, **321**, 862–864.

Bambach, R.K. (1977) Species richness in marine benthic habitats throughout the Phanerozoic. *Paleobiology*, **3**, 152–157.

Barsbold, R. & Osmolska, H. (1991) Ornithomimosauria. In: D.B. Weishampel, P. Dodson & H. Osmolska (eds) *The Dinosauria*, pp. 225–244. University of California Press, Berkeley.

Belyaev, G.M. (1990) Is it valid to isolate the genus *Xyloplax* as an independent class of echinoderms? *Zoologicheskii Zhurnal*, **69**, 83–96 (in Russian).

Benton, M.J. (1985) Mass extinction among non-marine tetrapods. *Nature*, **316**, 811–814.

Benton, M.J. (1986) More than one event in the late Triassic mass extinction. *Nature*, **321**, 857–861.

Benton, M.J. (1989) Mass extinction among tetrapods and the quality of the fossil record. *Philosophical Transactions of the Royal Society of London*, **B325**, 369–386.

Benton, M.J. (1990) Phylogeny of the major tetrapod groups: morphological data and divergence dates. *Journal of Molecular Evolution*, **30**, 409–424.

Bergström, J. (1989) The origin of animal phyla and the new phylum Procoelomata. *Lethaia*, **22**, 259–269.

Blake, D.B. (1987) A classification and phylogeney of post-Palaeozoic sea stars (Asteroidea: Echinodermata). *Journal of Natural History*, **21**, 481–528.

Bookstein, F.L. (1987) Random walks and the existence of evolutionary rates. *Paleobiology*, **13**, 446–464.

Bottjer, D.J. & Jablonski, D. (1988) Paleoenvironmental patterns in the evolution of post-Paleozoic benthic marine invertebrates. *Palaios*, **3**, 540–560.

Brady, R.H. (1985) On the independence of systematics. *Cladistics*, **1**, 113–126.

Bremer, K. (1990) Combinable component consensus. *Cladistics*, **6**, 369–372.

Bremer, K. (1992) Ancestral areas: a cladistic reinterpretation of the centers of origin concept. *Systematic Biology*, **41**, 436–445.

Briggs, D.E.G. & Gall, J.-C. (1990) The continuum in soft-bodied biotas from transitional environments: a quantitative comparison of Triassic and Carboniferous Konservat-Lägerstatten. *Paleobiology*, **16**, 204–218.

Briggs, D.E.G., Fortey, R.A. & Clarkson, E.N.K. (1988) Extinction and the fossil record of the arthropods. In: G.P. Larwood (ed.) *Extinction and Survival in the Fossil Record*, pp. 171–209. Systematics Association Special Volume, No. 34. Clarendon Press, Oxford.

Briggs, D.E.G., Fortey, R.A. & Wills, M.A. (1992) Morphological disparity in the Cambrian. *Science*, **256**, 1670–1673.

Budd, A.F. & Coates, A.G. (1992) Nonprogressive evolution in a clade of Cretaceous *Monastrea*-like corals. *Paleobiology*, **18**, 425–446.

Buzas, M.A., Koch, C.F., Culver, S.J. & Sohl, N.F. (1982) On the distribution of species occurrence. *Paleobiology*, **8**, 142–150.

Calloman, J.H. (1981) Dimorphism in ammonites. In: M.R. House & J.R. Senior (eds) *The Ammonoidea*, pp. 257–273. Systematics Association Special Volume, No. 18. Academic Press, London.

Campbell, K.S.W. & Marshall, C.R. (1986) Rates of evolution among Palaeozoic echinoderms. In: K.S.W. Campbell & M.F. Day (eds) *Rates of Evolution*, pp. 61–100. Allan & Unwin, London.

Carpenter, J.M. (1988) Choosing among multiple equally parsimonious cladograms. *Cladistics*, **4**, 291–296.

Carter, B.D. & McKinney, M.L. (1992) Eocene echinoids, the Suwannee Strait, and biogeographic taphonomy. *Paleobiology*, **18**, 299–325.

Carter, J.G. (ed.) (1990) *Skeletal Biomineralization: Patterns, Processes, and Evolutionary Trends*, Vols 1 & 2. Van Nostrand Reinhold, New York.

Cheetham, A.H. (1986) Tempo of evolution in a Neogene bryozoan: rates of morphometric change within and across species boundaries. *Paleobiology*, **12**, 190–202.

Cheetham, A.H. (1987) Tempo of evolution in a Neogene bryozoan: are trends in single morphological characters misleading? *Paleobiology*, **13**, 286–296.

Cheetham, A.H. & Hayek, L.C. (1988) Phylogeny reconstruction in the Neogene bryozoan *Metrarabdotos*: a paleontologic evaluation of methodology. *Historical Biology*, **1**, 65–83.

Cherry, L.M., Case, S.M., Kunkel, J.G., Wyles, J.S. & Wilson, A.C. (1982) Body shape metrics and organismal evolution. *Evolution*, **36**, 914–933.

Clark, C. & Curran, D.J. (1986) Outgroup analysis, homoplasy, and global parsimony: a response to Maddison, Donoghue, and Maddison. *Systematic Zoology*, **35**, 422–426.

Cloud, P.E. (1948) Some problems and patterns of evolution exemplified by fossil invertebrates. *Evolution*, **2**, 322–350.

Cloutier, R. (1991) Patterns, trends and rates of evolution within the Actinistia. *Environmental Biology of Fishes*, **32**, 23–58.

Cloutier, R. (Submitted) Extinct taxa and phylogeny reconstruction: selection of taxa and influence of anatomical incompleteness. *Systematic Biology*.

Cobban, W.A. (1969) The late Cretaceous ammonites *Scaphites leei* Reeside and *Scaphites hippocrepis* (De Kay) in the western interior of the United States. *US Geological Survey Professional Papers*, **619**, 10–29.

Cocks, L.R.M. (1988) Brachiopods across the Ordovician-Silurian boundary. In: L.R.M. Cocks (ed.) A global analysis of the Ordovician-Silurian boundary. *Bulletins of the British Museum (Natural History) (Geology)*, **43**, 311–315.

Coombs, W.P. & Maryanska, T. (1991) Ankylosauria. In: D.B. Weishampel, P. Dodson & H. Osmolska (eds) *The Dinosauria*, pp. 456–483. University of California Press, Berkeley.

Cooper, R.A. & Fortey, R.A.F. (1983) Development of the graptoloid rhabdosome. *Alcheringa*, **7**, 201–221.

Cooper, R.A. & Lindholm, K. (1990) A precise worldwide correlation of early Ordovician graptolite sequences. *Geological Magazine*, **127**, 497–525.

Cooper, R.A. & Ni, Y. (1986) Taxonomy, phylogeny and variability of *Pseudograptus* Beavis. *Palaeontology*, **29**, 313–363.

Cracraft, J. (1981) The use of functional and adaptive criteria in phylogenetic systematics. *American Zoologist*, **21**, 21–36.

Cracraft, J. (1982) A non-equilibrium theory for the rate-control of speciation and extinction and the origin of macroevolutionary patterns. *Systematic Zoology*, **31**, 348–365.

Cracraft, J. (1983) Species concepts and speciation analysis. *Current Ornithology*, **1**, 159–187.

Cracraft, J. (1984) Conceptual and methodological aspects of the study of evolutionary rates. In: N. Eldredge & S. Stanley (eds) *Living Fossils*, pp. 95–104. Springer-Verlag, New York.

Cracraft, J. (1989) Species as entities of biological theory. In: M. Ruse (ed.) *What the Philosophy of Biology is*, pp. 31–52. Kluwer Academic, Dordrecht.

Craske, A.J. & Jefferies, R.P.S. (1989) A new mitrate from the Upper Ordovician of Norway, and a new approach to subdividing a plesion. *Palaeontology*, **32**, 69–99.

Cripps, A.P. (1991) A cladistic analysis of the cornutes (stem chordates). *Zoological Journal of the Linnean Society*, **102**, 333–366.

Crisci, J.V. & Stuessy, T.F. (1980) Determining primitive character states for phylogenetic reconstruction. *Systematic Botany*, **5**, 112–135.

Cubitt, J.M. & Reyment, R.A. (1982) *Quantitative stratigraphical correlation*. Wiley & Sons, New York.

Culver, S.L., Buzas, M.A. & Collins, L.S. (1987) On the value of taxonomic standardization in evolutionary studies. *Paleobiology*, **13**, 169–176.

David, B. & Laurin, B. (1989) Déformation ontogénée et évolutive des organismes: l'approche par la méthode des points homologues. *Comptes Rendus, Académie des Sciences, Paris, série 2*, **309**, 1271–1276.

Davies, D.J., Powell, E.N. & Stanton, R.J. (1989) Relative rates of shell dissolution and net sediment accumulation – a commentary: can shell beds form by the gradual accumulation of biogenic debris on the sea floor? *Lethaia*, **22**, 207–212.

Debry, R.W. (1992) The consistency of several phylogenetic-inference methods under varying evolutionary rates. *Molecular Biology and Evolution*, **9**, 537–551.

de Pinna, M.C.C. (1991) Concepts and tests of homology in the cladistic paradigm. *Cladistics*, **7**, 367–394.

de Queiroz, K. & Donoghue, M.J. (1988) Phylogenetic systematics and the species problem. *Cladistics*, **4**, 317–338.

de Queiroz, K. & Donoghue, M.J. (1990a) Phylogenetic systematics or Nelson's version of cladistics? *Cladistics*, **6**, 61–75.

de Queiroz, K. & Donoghue, M.J. (1990b) Phylogenetic systematics and species revisited. *Cladistics*, **6**, 83–90.

de Queiroz, K. & Gauthier, J. (1990) Phylogeny as a central principle in taxonomy: phylogenetic definitions of taxon names. *Systematic Zoology*, **39**, 307–322.

de Queiroz, K. & Gauthier, J. (1992) Phylogenetic taxonomy. *Annual Reviews in Ecology and Systematics*, **23**, 449–480.

Dial, K.P. & Marzluff, J.M. (1989) Nonrandom diversification within taxonomic assemblages. *Systemic Zoology*, **38**, 26–37.

Dingus, L. (1984) Effects of stratigraphic completeness on interpretations of extinction rates across the Cretaceous-Tertiary boundary. *Paleobiology*, **10**, 420–438.

Donoghue, M.J. (1985) A critique of the biological species concept and recommendations for a phylogenetic alternative. *Bryologist*, **88**, 172–181.

Donoghue, M.J., Doyle, J., Gauthier, J., Kluge, A. & Rowe, T. (1989) The importance of fossils in phylogeny reconstruction. *Annual Reviews in Ecology and Systematics*, **20**, 431–460.

Donovan, S.K. (1991) Echinoderm taphonomy. In: S.K. Donovan (ed.) *The Processes of Fossilization*, pp. 241–269. Belhaven Press, London.

Dowsett, H.J. (1989) Application of graphic correlation method to Pliocene marine sequences. *Marine Micropaleontology*, **14**, 3–32.

Doyle, J. & Donoghue, M.J. (1987) The importance of fossils in elucidating seed plant phylogeny and macroevolution. *Reviews in Paleobotany and Palynology*, **50**, 63–95.

Doyle, J. & Donoghue, M.J. (1993) Phylogenies and angiosperm diversification. *Paleobiology*, **19**, 141–167.

Doyle, P. (1985) Sexual dimorphism in the belemnite *Youngibelus* from the Lower Jurassic of Yorkshire. *Palaeontology*, **28**, 133–146.

Dullo, W.C. & Bandel, K. (1988) Diagenesis of molluscan shells: a case study from cephalopods. In: J. Wiedmann & J. Kullmann (eds) *Second International Cephalopod Symposium: Cephalopods Present and Past*, pp. 719–729. Schweizerbart, Stuttgart.

Edgecombe, G.D. (1992) Trilobite phylogeny and the Cambrian-Ordovician 'event': cladistic reappraisal. In: M.J. Novacek & Q.D. Wheeler (eds) *Extinction and Phylogeny*, pp. 144–177. Columbia University Press, New York.

Edwards, L.E. (1989) Supplemented graphic correlation: a powerful tool for paleontologists and nonpaleontologists. *Palaios*, **4**, 127–143.

Edwards, L.E. (1991) Quantitative biostratigraphy. In: N.L. Gilinsky & P.W. Signor (eds) Analytical paleobiology. *Short Courses in Paleontology*, **4**, 39–58.

Eernisse, D.J. & Kluge, A.G. (1993) Taxonomic congruence versus total evidence, and amniote phylogeny inferred from fossils, molecules, and morphology. *Molecular Biology and Evolution*, **10**, 1170–1195.

Eldredge, N. (1979) Cladism and common sense. In: J. Cracraft & N. Eldredge (eds) *Phylogenetic Analysis and Paleontology*, pp. 165–198. Columbia University Press, New York.

Eldredge, N. (1985) *Unfinished Synthesis*. Oxford University Press, Oxford.

Eldredge, N. (1986) Information, economics, and evolution. *Annual Reviews in Ecology and Systematics*, **17**, 351–369.

Eldredge, N. & Cracraft, J. (1980) *Phylogenetic Patterns and the Evolutionary Process*. Columbia University Press, New York.

Eldredge, N. & Salthe, S.N. (1984) Hierarchy and evolution. *Oxford Surveys in Evolutionary Biology*, **1**, 184–208.

Engelmann, G.F. & Wiley, E.O. (1977) The place of ancestor–descendant relationships in phylogeny reconstruction. *Systematic Zoology*, **26**, 1–11.

Evander, R.L. (1989) Phylogeny of the family Equidae. In: D.R. Prothero & R.M. Schoch (eds) *The Evolution of Perissodactyls*, pp. 109–126. Clarendon Press, Oxford.

Farris, J.S. (1969) A successive approximation approach to character weighting. *Systematic Zoology*, **18**, 374–385.

Farris, J.S. (1974) Formal definitions of paraphyly and polyphyly. *Systematic Zoology*, **23**, 548–554.

Farris, J.S. (1977) On the phenetic approach to vertebrate classification. In: M.K. Hecht, P.C. Goody & B.M. Hecht (eds) *Major Patterns in Vertebrate Evolution*, pp. 823–850. Plenum Press, New York.

Farris, J.S. (1979) The information content of the phylogenetic system. *Systematic Zoology*, **28**, 483–519.

Farris, J.S. (1980) The efficient diagnoses of the phylogenetic system. *Systematic Zoology*, **29**, 386–401.

Farris, J.S. (1981) Distance data in phylogenetic analysis. In: V.A. Funk & D.R. Brooks (eds) *Advances in Cladistics: Proceedings of the First Meeting of the*

Willi Hennig Society, pp. 7–36. New York Botanic Garden, New York.

Farris, J.S. (1982a) Simplicity and informativeness in systematics and phylogeny. *Systematic Zoology*, **31**, 413–444.

Farris, J.S. (1982b) Outgroups and parsimony. *Systematic Zoology*, **31**, 328–334.

Farris, J.S. (1983) The logical basis of phylogenetic analysis. In: N.I. Platnick & V.A. Funk (eds) *Advances in Cladistics*, Vol. 2, pp. 7–36. Columbia University Press, New York.

Farris, J.S. (1988) *HENNIG86, Version 1.5*. Port Jefferson Station, New York. (Computer program.)

Farris, J.S. (1989) The retention index and rescaled consistency index. *Cladistics*, **5**, 417–419.

Farris, J.S. (1991) Hennig defined paraphyly. *Cladistics*, **7**, 297–304.

Felsenstein, J. (1978) Cases in which parsimony or compatibility methods will be positively misleading. *Systematic Zoology*, **27**, 401–410.

Felsenstein, J. (1988) Phylogenies from molecular sequences: inference and reliability. *Annual Reviews in Genetics*, **22**, 521–565.

Felsenstein, J. & Kishino, H. (1993) Is there something wrong with the bootstrap on phylogenies? A reply to Hillis & Bull. *Systematic Biology*, **42**, 193–200.

Fisher, D.C. (1991) Phylogenetic analysis and its application in evolutionary paleobiology. In: N.L. Gilinsky & P.W. Signor (eds) Analytical paleobiology. *Short Courses in Paleontology*, **4**, 103–122.

Flessa, K.W. & Brown, T.J. (1983) Selective solution of macroinvertebrate calcareous hard parts: a laboratory study. *Lethaia*, **16**, 193–205.

Flessa, K.W. & Jablonski, D. (1983) Extinction is here to stay. *Paleobiology*, **9**, 315–321.

Flessa, K.W. & Jablonski, D. (1985) Declining Phanerozoic background extinction rates: effects of taxonomic structure? *Nature*, **313**, 216–218.

Flessa, K.W. & Levinton, J.S. (1975) Phanerozoic diversity patterns: tests for randomness. *Journal of Geology*, **83**, 239–248.

Foote, M. (1988) Survivorship analysis of Cambrian and Ordovician trilobites. *Paleobiology*, **14**, 258–271.

Foote, M. (1991a) Morphological patterns of diversification: examples from trilobites. *Palaeontology*, **34**, 461–485.

Foote, M. (1991b) Morphological and taxonomic diversity in a clade's history: the blastoid record and stochastic simulations. *Contibutions to the Museum of Paleontology, University of Michigan*, **28**, 101–140.

Foote, M. (1991c) Analysis of morphological data. In: N.L. Gilinsky & P.W. Signor (eds) Analytical paleobiology. *Shourt Courses in Paleontology*, **4**, 59–86.

Foote, M. (1992a) Paleozoic record of morphological diversity in blastozoan echinoderms. *Proceedings of the National Academy of Sciences, USA*, **89**, 7325–7329.

Foote, M. (1992b) Rarefaction analysis of morphological and taxonomic diversity. *Paleobiology*, **18**, 1–16.

Foote, M. & Gould, S.J. (1992) Cambrian and Recent morphological disparity. *Science*, **258**, 1816.

Fordham, B.G. (1986) Miocene–Pleistocene planktic foraminifers from D.S.D.P. sites 208 and 77, and phylogeny and classification of Cenozoic species. *Evolutionary Monographs*, **6**, 1–200.

Forey, P.L. (1991) *Latimeria chalumnae* and its pedigree. *Environmental Biology of Fishes*, **32**, 75–97.

Forey, P.L., Humphries, C.J., Kitching, I.J., Scotland, R.W., Siebert, D.J. & Williams, D.M. (1992) *Cladistics: a Practical Course in Systematics*. Oxford University Press, Oxford.

Fortey, R.A. (1983) Cambro-Ordovician trilobites from the boundary beds in western Newfoundland and their phylogenetic significance. *Special Papers in Palaeontology*, **30**, 179–212.

Fortey, R.A. (1985) Gradualism and punctuated equilibria as competing and complementary theories. *Special Papers in Palaeontology*, **33**, 17–28.

Fortey, R.A. (1988) Seeing is believing: gradualism and punctuated equilibria in the fossil record. *Science Progress, Oxford*, **72**, 1–19.

Fortey, R.A. (1989) There are extinctions and extinctions: examples from the Lower Palaeozoic. *Philosophical Transactions of the Royal Society, London*, **B325**, 327–355.

Fortey, R.A. (1990a) Ontogeny, hypostome attachment and trilobite classification. *Palaeontology*, **33**, 529–576.

Fortey, R.A. (1990b) Trilobite evolution and systematics. In: S.J. Culver (ed.) Arthropod paleobiology. *Short Courses in Paleontology*, **3**, 44–65.

Fortey, R.A. & Chatterton, B.D.E. (1988) Classification of the trilobite suborder Asaphina. *Palaeontology*, **31**, 165–222.

Fortey, R.A. & Cooper, R.A. (1986) A phylogenetic classification of the graptoloids. *Palaeontology*, **29**, 631–654.

Fortey, R.A. & Jefferies, R.P.S. (1982) Fossils and phylogeny – a compromise approach. In: K.A. Joysey & A.E. Friday (eds) *Problems of Phylogenetic Reconstruction*, pp. 197–234. Systematics Association Special Volume, No. 21. Academic Press, London.

Fortey, R.A. & Owens, R.M. (1991) Evolutionary radiations in the Trilobita. In: P.D. Taylor & G.P. Larwood (eds) *Major Evolutionary Radiations*, pp. 139–164. Systematics Association Special Volume, No. 42. Oxford University Press, Oxford.

Fortey, R.A. & Whittington, H.B. (1989) The Trilobita as a natural group. *Historical Biology*, **2**, 125–138.

Fox, W.T. (1987) Harmonic analysis of periodic extinctions. *Paleobiology*, **13**, 257–271.

Funk, V.A. & Brooks, D.R. (1990) *Phylogenetic Systematics as the Basis of Comparative Biology*. Smithsonian Institution Press, Washington, D.C.

Garstang, W. (1928) The morphology of the Tunicata and its bearing on the phylogeny of the Chordata. *Quarterly Journal of Microscopical Science*, **72**, 51–187.

Gauthier, J., Kluge, A. & Rowe, T. (1988) Amniote phylogeny and the importance of fossils. *Cladistics*, **4**, 105–209.

Geary, D.H. (1990) Patterns of evolutionary tempo and mode in the radiation of *Melanopsis* (Gastropoda; Melanopsidae). *Paleobiology*, **16**, 492–511.

Geary, D.H. (1992) An unusual pattern of divergence between two fossil gastropods: ecophenotypy, dimorphism, or hybridization? *Paleobiology*, **18**, 93–109.

Ghiselin, M.T. (1984a) 'Definition', 'character', and other equivocal terms. *Systematic Zoology*, **33**, 104–110.

Ghiselin, M.T. (1984b) Narrow approaches to phylogeny: a review of nine books on cladism. In: R. Dawkins & M. Ridley (eds) *Oxford Surveys in Evolutionary Biology 1*, pp. 209–222. Oxford University Press, Oxford.

Gilinsky, N.L. (1988) Survivorship in the Bivalvia: comparing living and extinct genera and families. *Paleobiology*, **14**, 370–386.

Gilinsky, N.L. & Bambach, R.K. (1987) Asymmetrical patterns of origination and extinction in higher taxa. *Paleobiology*, **13**, 427–445.

Gilinsky, N.L. & Good, I.J. (1991) Probabilities of origination, persistence, and extinction of families of marine invertebrate life. *Paleobiology*, **17**, 145–166.

Gingerich, P.D. (1979) The stratigraphic approach to phylogeny reconstruction in vertebrate paleontology. In: J. Cracraft & N. Eldredge (eds) *Phylogenetic Analysis and Paleontology*, pp. 41–77. Columbia University Press, New York.

Gingerich, P.D. (1983) Rates of evolution: effects of time and temporal scaling. *Science*, **222**, 159–161.

Gingerich, P.D. & Schoeninger, M. (1977) The fossil record and primate phylogeny. *Journal of Human Evolution*, **6**, 484–505.

Gould, S.J. (1984) Smooth curve of evolutionary rate: a psychological and mathematical artefact. *Science*, **226**, 994–996.

Gould, S.J. (1991) The disparity of the Burgess Shale arthropod fauna and the limits of cladistic analysis: why we must strive to quantify morphospace. *Paleobiology*, **17**, 411–423.

Gould, S.J., Raup, D.M., Sepkoski, J.J., Schopf, T.J.M. & Simberloff, D.S. (1977) The shape of evolution: a comparison of real and random clades. *Paleobiology*, **3**, 23–40.

Gould, S.J., Gilinsky, N.L. & German, R.Z. (1987) Asymmetry of lineages and the direction of evolutionary time. *Science*, **236**, 1437–1441.

Gradstein, F.P., Agterberg, F.P., Brower, J.C. & Schwarzacher, W.S. (1985) *Quantitative Stratigraphy*. Reidel, Dordrecht.

Hallam, A. (1989) The case for sea-level change as a dominant causal factor in mass extinction of marine invertebrates. *Philosophical Transactions of the Royal Society, London*, **B325**, 437–455.

Haq, B.U., Hardenbol, J. & Vail, P.R. (1987) Chronology of fluctuating sea levels since the Triassic. *Science*, **235**, 1156–1166.

Harland, W.B., Holland, C.H., House M.R., Hughes, N.F., Reynolds, A.B., Rudwick, M.J.S., Satterwaite, G.E., Tarlo, L.B.H. & Willey, E.C. (eds) (1967) *The Fossil Record*. Geological Society, London.

Harper, C.W. (1976) Phylogenetic inference in paleontology. *Journal of Paleontology*, **50**, 180–193.

Harper, C.W. (1980) Relative age inference in palaeontology. *Lethaia*, **13**, 239–248.

Harrison, R.G. (1991) Molecular changes at speciation. *Annual Reviews in Ecology and Systematics*, **22**, 281–308.

Hart, M.B. (1990) Major evolutionary radiations of the planktonic Foraminiferida. In: P.D. Taylor & G.P. Larwood (eds) *Major Evolutionary Radiations*. Systematics Association Special Volume, No. 42, pp. 59–72. Clarendon Press, Oxford.

Hauser, D.L. (1992) Similarity, falsification and character state order – a reply to Wilkinson. *Cladistics*, **8**, 339–344.

Hauser, D.L. & Presch, W. (1991) The effect of ordered characters on phylogenetic reconstruction. *Cladistics*, **7**, 243–266.

Hay, W.W. (1972) Probablistic stratigraphy. *Ecologae Geologicae Helvetiae*, **65**, 255–266.

Hennig, W. (1966) *Phylogenetic Systematics*. University of Illinois Press, Urbana.

Hennig, W. (1969) *Die Stammesgeschichte der Insekten*. Kramer, Frankfurt.

Hennig, W. (1981) *Insect Phylogeny*. John Wiley, New York.

Hillis, D.M. (1991) Discrimination between phylogenetic signal and random noise in DNA sequences. In: M.M. Myamoto & J. Cracraft (eds) *Phylogenetic Analysis of DNA Sequences*, pp. 278–294. Oxford University Press, Oxford.

Hillis, D.M. & Bull, J.J. (1993) An empirical test of bootstrapping as a method for assessing confidence in phylogenetic analysis. *Systematic Biology*, **42**, 182–192.

Hillis, D.M. & Huelsenbeck, J.P. (1992) Signal, noise and reliability in molecular phylogenetic analysis. *Heredity*, **83**, 189–195.

Hillis, D.M. & Moritz, C. (eds) (1989) *Molecular Systematics*. Sinauer Associates, Sunderland, Massachusetts.

Hoffman, A. (1989) *Arguments on Evolution: a Paleontologist's Perspective*. Oxford University Press, Oxford.

Hoffman, A. & Kitchell, J.A. (1984) Evolution in a pelagic planktic system: a paleobiologic test of models of multispecies evolution. *Paleobiology*, **10**, 9–33.

Hohenegger, J. & Tatzreiter, F. (1992) Morphometric methods in determination of ammonite species. exemplified through *Balatonites* shells (Middle Triassic). *Journal of Paleontology*, **66**, 801–816.

Holman, E.W. (1989) Some evolutionary correlates of higher taxa. *Paleobiology*, **15**, 357–363.

Howarth, M.K. (1991) The ammonite family Hildoceratidae in the Lower Jurassic of Britain, Part 1. *Monographs of the Palaeontographical Society*, **145** (586), 1–106, plates 1–16.

Huelsenbeck, J.P. (1991a) Tree-length distribution skewness: an indicator of phylogenetic information. *Systematic Zoology*, **40**, 257–270.

Huelsenbeck, J.P. (1991b) When are fossils better than extant taxa in phylogenetic analysis? *Systematic Zoology*, **40**, 458–469.

Hull, D.L. (1970) Contemporary systematic philosophies. *Annual Review of Systematics and Evolution*, **1**, 19–54.

Humphries, C.J. & Parenti, L.R. (1986) *Cladistic Biogeography*. Oxford Monographs in Biogeography, No. 2. Clarendon Press, Oxford.

Jablonski, D. (1986a) Larval ecology and macroevolution in marine invertebrates. *Bulletin of Marine Sciences*, **39**, 568–587.

Jablonski, D. (1986b) Background and mass extinctions: the alternation of macroevolutionary regimes. *Science*, **231**, 129–133.

Jablonski, D. (1987) Heritability at the species level: analysis of geographic ranges of Cretaceous mollusks. *Science*, **238**, 360–363.

Jablonski, D. (1988) Estimates of species duration: response to Russell & Lindberg. *Science*, **240**, 969.

Jablonski, D. & Bottjer, D.J. (1990a) The origin and diversification of major groups: environmental patterns and macroevolutionary lags. In: P.D. Taylor & G.P. Larwood (eds) *Major Evolutionary Radiations*. Systematics Association Special Volume number 42, pp. 17–57. Clarendon Press, Oxford.

Jablonski, D. & Bottjer, D.J. (1990b) Onshore–offshore trends in marine invertebrate evolution. In: R.M. Ross & W.D. Allmon (eds) *Causes of Evolution: a Paleontological Perspective*, pp. 21–75. University of Chicago Press, Chicago.

Jablonski, D. & Bottjer, D.M. (1991) Environmental patterns in the origins of higher taxa: the post-Paleozoic fossil record. *Science*, **252**, 1831–1833.

Jablonski, D. & Smith, A.B. (1990) Ecology and phylogeny: environmental patterns in the evolution of the echinoid order Salenioida. *Geological Society of America Abstracts with Programs*, **22**, A266.

Jablonski, D., Sepkoski, J.J., Bottjer, D. & Sheehan, P.M. (1983) Onshore–offshore patterns in the evolution of Phanerozoic shelf communities. *Science*, **222**, 1123–1124.

Jackson, J.B.C. & Cheetham, A.H. (1990) Evolutionary significance of morpho-species: a test with cheilostome Bryozoa. *Science*, **248**, 579–583.

Jefferies, R.S.P. (1979) The origin of the chordates – a methodological essay. In: M.R. House (ed.) *The Origin of Major Invertebrate Groups*. Systematics Association Special Volume, No. 12.

Jefferies, R.S.P. (1986) *The Origin of the Vertebrates*. British Museum (Natural History), London.

Jensen, J.S. (1990) Plausibility and testibility: asssessing in consequences of evolutionary innovation. In: M.H. Niteki (ed.) *Evolutionary Novelties*, pp. 171–190. University of Chicago Press, Chicago.

Jensen, M. (1981) Morphology and classification of Euechinoidea Bronn, 1860; a cladistic analysis. *Videnskabelige Meddelelser fra dansk naturhistorisk Forening i Kjøbenhavn*, **143**, 7–99.

Jones, D.S. & Nicol, D. (1986) Origination, survivorship, and extinction of rudist taxa. *Journal of Paleontology*, **60**, 107–115.

Kidell, S.M. & Baumiller, T. (1990) Experimental disintegration of regular echinoids: roles of temperature, oxygen and decay thresholds. *Paleobiology*, **16**, 247–272.

Kitchell, J.A. & Carr, T.R. (1985) Nonequilibrium model of diversification: faunal turnover dynamics. In: J.W. Valentine (ed.) *Phanerozoic Diversity Patterns: Profiles in Macroevolution*, pp. 277–309. Princeton University Press, Princeton.

Kitchell, J.A. & MacLeod, N. (1988) Macroevolutionary interpretations of symmetry and synchroneity in the fossil record. *Science*, **240**, 1190–1193.

Kitchell, J.A., Estabrook, G. & MacLeod, N. (1987) Testing for equality of evolutionary rates. *Paleobiology*, **13**, 272–285.

Kluge, A.G. (1984) The relevance of parsimony to phylogenetic inference. In: T. Duncan & T.F. Stuessey (eds) *Cladistics: Perspectives on the Reconstruction of Evolutionary History*, pp. 24–38. Columbia University Press, New York.

Kluge, A.G. (1989) A concern for evidence and a phylogenetic hypothesis of relationships among *Epicrates* (Boidae, Serpentes). *Systematic Zoology*, **38**, 7–25.

Kluge, A.G. & Wolf, A.J. (1993) Cladistics: what's in a word? *Cladistics*, **9**, 183–200.

Knowlton, N. (1993) Sibling species in the sea. *Annual Reviews of Ecology and Systematics*, **24**, 189–216.

Koch, C.F. (1987) Prediction of sample size effects on the measured temporal and geographic distribution patterns of species. *Paleobiology*, **13**, 100–107.

Koch, C.F. (1991) Sampling from the fossil record. In: N.L. Gilinsky & P.W. Signor (eds) Analytical paleobiology. *Short Courses in Paleontology*, **4**, 4–18.

Koch, C.F. & Sohl, N.F.P. (1983) Preservational effects in paleoecological studies: Cretaceous mollusc examples. *Paleobiology*, **9**, 26–34.

Landman, N.H. (1989) Iterative progenesis in Upper Cretaceous ammonites. *Paleobiology*, **15**, 95–117.

Lanyon, S.M. (1987) Jackknifing and bootstrapping: important 'new' statistical techniques for ornithologists. *The Auk*, **104**, 144–145.

Lanyon, S.M. (1988) The stochastic mode of molecular evolution: what consequences for systematic investigations? *The Auk*, **105**, 565–573.

Laurin, B. & David, B. (1990a) Analysis of shape change among brachiopods by using landmarks: application to the Jurassic Septaliphora lineage. In: D.I. MacKinnon, D.E. Lee & J.D. Campbell (eds) *Brachiopods Through Time*, pp. 81–87. Balkema, Rotterdam.

Laurin, B. & David, B. (1990b) Mapping morphological changes in the spatangoid *Echinocardium*: applications to ontogeny and interspecific comparisons. In: C. de Ridder, P. Dubois, M.C. Lahaye & M. Jangoux (eds) *Echinoderm Research*, pp. 131–136. Balkema, Rotterdam.

Lauterbach, K.E. (1980) Schlüssereignisse in der Evolution des Grundplans der Arachnata (Arthropoda). *Abhandlungen des Naturwissenschaftlichen Vereins in Hamburg*, **23**, 163–327.

Lazarus, D. (1986) Tempo and mode of morphologic evolution near the origin of the radiolarian lineage *Pterocanium prismatium*. *Paleobiology*, **12**, 175–189.

Lehmann, U. (1981) *The Ammonites: Their Life and Their World*. Cambridge University Press, Cambridge.

Levinton, J. (1988) *Genetics, Paleontology and Macroevolution*. Cambridge University Press, Cambridge.

Levinton, J.S. & Farris, J.S. (1987) On the estimation of taxonomic longevity from Lyellian curves. *Paleobiology*, **13**, 479–483.

Lipscomb, D.L. (1992) Parsimony, homology and the analysis of multistate characters. *Cladistics*, **8**, 45–65.

McCune, A.R. (1986) A revision of *Semionotus* (Pisces: Semionotidae) from the Triassic and Jurassic of Europe. *Palaeontology*, **29**, 213–233.

McKinney, M.L. (1991) Completeness of the fossil record: an overview. In: S.K. Donovan (ed.) *The Processes of Fossilization*, pp. 66–83. Belhaven Press, London.

McKinney, M.L. & Oyen, C.W. (1989) Causation and nonrandomness in biological and geological time series: temperature as a proximal control of extinction and diversity. *Palaios*, **4**, 3–15.

MacLeod, N. (1991) Punctuated anagenesis and the importance of stratigraphy to paleobiology. *Paleobiology*, **17**, 167–188.

MacLeod, N. & Keller, G. (1991) How complete are Cretaceous/Tertiary boundary sections? A chronostratigraphic estimate based on graphic correlation. *Geological Society of America Bulletin*, **103**, 1439–1457.

Maddison, W.P. & Maddison, D.R. (1992) *MacCLADE: Analysis of Phylogeny and Character Evolution, Version 3.0*. Sinauer Associates, Sunderland, Massachusetts. (Computer program and manual.)

Maddison, W.P., Donoghue, M.J. & Maddison, D.R. (1984) Outgroup analysis and parsimony. *Systematic Zoology*, **33**, 83–103.

Malmgren, B.A., Berggren, W.A. & Lohmann, G.P. (1983) Evidence for punctuated gradualism in the late Neogene *Globorotalia tumida* lineage of planktonic foraminifera. *Paleobiology*, **9**, 377–389.

Margush, T. & McMorris, F.R. (1981) Consensus n-trees. *Bulletins of Mathematical Biology*, **43**, 239–244.

Marshall, C.R. (1990) Confidence intervals on stratigraphic ranges. *Paleobiology*, **16**, 1–10.

Marshall, C.R. (1991) Estimation of taxonomic ranges from the fossil record. In: N.L. Gilinsky & P.W. Signor (eds) Analytical paleobiology. *Short Courses in Paleontology*, **4**, 19–38.

Marshall, C.R. & Schultze, H.-P. (1992) Relative importance of molecular, neontological, and paleontological data in understanding the biology of the vertebrate invasion of land. *Journal of Molecular Evolution*, **35**, 93–101.

Maxwell, W.D. & Benton, M.J. (1990) Historical tests of the absolute completeness of the fossil record of tetrapods. *Paleobiology*, **16**, 322–335.

Mayr, E. (1969) *Principles of Systematic Zoology*. McGraw-Hill, New York.

Mayr, E. (1976) *Evolution and the Diversity of Life: Selected Essays*. Harvard Univesity Press, Cambridge, Massachusetts.

Mayr, E. (1982) *The Growth of Biological Thought: Diversity, Evolution and Inheritance*. Harvard University Press, Cambridge, Massachussetts.

Meyer, A. & Dolven, S.I. (1992) Molecules, fossils and the origin of tetrapods. *Journal of Molecular Evolution*, **35**, 102–113.

Meyer, A. & Wilson, A.C. (1990) Origin of tetrapods inferred from their mitochondrial DNA affiliation to lungfish. *Journal of Molecular Evolution*, **31**, 359–364.

Michaux, B. (1989) Cladograms can reconstruct phylogenies: an example from the fossil record. *Alcheringa*, **13**, 21–36.

Mickevich, M.F. (1982) Transformation series analysis. *Systematic Zoology*, **31**, 461–478.

Mickevich, M.F. & Platnick, N.I. (1989) On the information content of classifications. *Cladistics*, **5**, 33–47.

Milner, A.R. (1990) The radiations of temnospondyl amphibians. In: P.D. Taylor & G.P. Larwood (eds) *Major Evolutionary Radiations*, pp. 321–349. Systematics Association Special Volume number 42. Clarendon Press, Oxford.

Mischler, B.D. & Brandon, R.N. (1987) Individuality, pleuralism, and the phylogenetic species concept. *Biological Philosophy*, **2**, 397–414.

Mischler, B.D. & Donoghue, M.J. (1982) Species concepts: a case for pleuralism. *Systematic Zoology*, **31**, 491–503.

Miyamoto, M.M. (1985) Consensus cladograms and general classifications. *Cladistics*, **1**, 186–189.

Miyamoto, M.M. & Cracraft, J. (1991) *Phylogenetic Analysis of DNA Sequences*. Oxford University Press, Oxford.

Mizuno, Y. (1991) Fossil echinoderms from the early Miocene Morozaki Group in the Chita peninsula, Central Japan. In: T. Yanagisawa, I. Yasumasu, C. Oguro, N. Suzuki & T. Motokawa (eds) *Biology of Echinoderms*, p. 532. Balkema, Rotterdam.

Mooi, R. (1990) Paedomorphosis, Aristotle's lantern, and the origin of the sand dollars (Echinodermata: Clypeasterioda). *Paleobiology*, **16**, 25–48.

Moore, R.C., Teichert, C. & Robison, R.A. (eds) (1953–86) *Treatise on Invertebrate Paleontology*. Geological Society of America and University of Kansas Press, Lawrence, Kansas.

Nee, S., Mooers, A.O. & Harvey, P.H. (1992) Tempo and mode of evolution revealed from molecular phylogenies. *Proceedings of the National Academy of Sciences, USA*, **89**, 8322–8326.

Nei, M. (1991) Relative efficiencies of different tree-making methods for molecular data. In: M.M. Myamoto & J. Cracraft (eds) *Phylogenetic Analysis of DNA Sequences*, pp. 90–128. Oxford University Press, Oxford.

Nelson, G. (1978a) From Candolle to Croizat: comments on the history of biogeography. *Journal of Historical Biology*, **11**, 269–305.

Nelson, G. (1978b) Ontogeny, phylogeny, paleontology, and the biogenic law. *Systematic Zoology*, **27**, 324–345.

Nelson, G. (1989a) Cladistics and evolutionary models. *Cladistics*, **5**, 275–289.

Nelson, G. (1989b) Species and taxa: systematics and evolution. In: D. Otto & J.A. Endler (eds) *Speciation and its Consequences*, pp. 60–81. Sinauer Associates, Sunderland, Massachusetts.

Nelson, G. & Platnick, N.I. (1981) *Systematics and Biogeography: Cladistics and Vicariance*. Columbia University Press, New York.

Newell, N.D. (1952) Periodicity in invertebrate evolution. *Journal of Paleontology*, **26**, 371–385.

Newell, N.D. (1967) Revolutions in the history of life. In: C.C. Albritton (ed.) Uniformity and simplicity: a symposium on the principle of the uniformity of nature. *Geological Society of America Special Paper*, **89**, 63–91.

Nixon, K.C. & Davis, J.I. (1991) Polymorphic taxa, missing values and cladistic analysis. *Cladistics*, **7**, 233–242.

Nixon, K.C. & Wheeler, Q.D. (1990) An amplification of the phylogenetic species concept. *Cladistics*, **6**, 211–223.

Norell, M.A. (1992) Taxic origin and temporal diversity: the effect of phylogeny. In: M.J. Novacek & Q.D. Wheeler (eds) *Extinction and Phylogeny*, pp. 88–118. Columbia University Press, New York.

Norell, M.A. & Novacek, M.J. (1992a) The fossil record and evolution: comparing cladistic and paleontologic evidence for vertebrate history. *Science*, **255**, 1690–1693.

Norell, M.A. & Novacek, M.J. (1992b) Congruence between superpositional and phylogenetic patterns: comparing cladistic patterns with fossil records. *Cladistics*, **8**, 319–338.

Novacek, M.J. (1989) Higher mammal phylogeny: the morphological-molecular synthesis. In: B. Fernholm, K. Bremer & H. Jornvall (eds) *The Hierarchy of Life*, pp. 421–435. Excerpta Medica, Amsterdam.

Novacek, M.J. (1992a) Fossils, topologies, missing data, and the higher level phylogeny of eutherian mammals. *Systematic Biology*, **41**, 58–73.

Novacek, M.J. (1992b) Fossils as critical data for phylogeny. In: M.J. Novacek & Q.D. Wheeler (eds) *Extinction and Phylogeny*, pp. 46–88. Columbia University Press, New York.

Novacek, M.J. & Norell, M.A. (1982) Fossils, phylogenies and taxonomic rates of evolution. *Systematic Zoology*, **31**, 366–375.

Novacek, M.J. & Wheeler, Q.D. (1992) Introduction: extinct taxa. In: M.J. Novacek & Q.D. Wheeler (eds) *Extinction and Phylogeny*, pp. 1–16. Columbia University Press, New York.

Oosterbrook, P. (1987) More appropriate definitions of paraphyly and polyphyly, with a comment on the Farris, 1974, model. *Systematic Zoology*, **36**, 103–108.

Padian, K. (1989) The whole real guts of evolution? *Paleobiology*, **15**, 73–78. (Review of J. Levinton, 1988.)

Palmer, A.R. (1965) Biomere – a new kind of biostratigraphic unit. *Journal of Paleontology*, **39**, 149–153.

Pandolfi, J.M. (1989) Phylogenetic analysis of the early tabulate corals. *Palaeontology*, **32**, 745–764.

Parsons, K.M. & Brett, C.E. (1991) Taphonomic processes and biases in modern marine environments: an actualistic perspective on fossil assemblage preservation. In: S.K. Donovan (ed.) *The Processes of Fossilization*, pp. 22–65. Belhaven Press, London.

Paterson, H.E.H. (1981) The continuing search for the unknown and unknowable: a critique of contemporary ideas on speciation. *South African Journal of Science*, **77**, 113–119.

Paterson, H.E.H. (1985) The recognition concept of species. In: E.S. Vrba (ed.) Species and speciation. *Transvaal Museum Monographs*, **4**, 21–29.

Patterson, C. (1977) The contribution of paleontology to teleostean phylogeny. In: M.K. Hecht, P.C. Goody & B.M. Hecht (eds) *Major Patterns in Vertebrate Evolution*, pp. 579–643. Plenum, New York.

Patterson, C. (1980) Cladistics. *Biologist*, **27**, 234–240.

Patterson, C. (1981) Significance of fossils in determining evolutionary relationships. *Annual Reviews in Ecology and Systematics*, **12**, 195–223.

Patterson, C. (1982) Morphological characters and homology. In: K.A. Joysey & A.E. Friday (eds) *Problems of Phylogenetic Reconstruction*, pp. 21–74. Systematics Association Special Volume, No. 21. Academic Press, London.

Patterson, C. (1983) How does phylogeny differ from ontogeny? In: B.C. Goodwin, H. Holder & C.C. Wylie (eds) *Development and Evolution*, pp. 1–31. Cambridge University Press, Cambridge.

Patterson, C. & Rosen, D.E. (1977) Review of ichthyodectiform and other Mesozoic teleost fishes and the theory and practice of classifying fossils. *Bulletins of the American Museum of Natural History*, **158**, 81–172.

Patterson, C. & Smith, A.B. (1987) Is the periodicity of extinction a taxonomic artefact? *Nature*, **330**, 248–251.

Patterson, C. & Smith, A.B. (1988) Periodicity in extinction: the role of systematics. *Ecology*, **70**, 902–811.

Paul, C.R.C. (1979) Early echinoderm radiation. In: M.R. House (ed.) *The Origin of Major Invertebrate Groups*, pp. 415–434. Systematics Association Special Volume, No. 12. Academic Press, London.

Paul, C.R.C. (1982) The adequacy of the fossil record. In: K.A. Joysey & A.E. Friday (eds) *Problems of Phylogenetic Reconstruction*, pp. 75–117. Academic Press, London.

Paul, C.R.C. (1985) The adequacy of the fossil record reconsidered. In: J.C.W. Cope & P.W. Skelton (eds) Evolutionary case histories from the fossil record. *Special Papers in Palaeontology*, **33**, 7–16.

Pearson, P.N. (1992) Survivorship analysis of fossil taxa when real-time extinction rates vary: the Paleogene planktonic foraminifera. *Paleobiology*, **18**, 115–131.

Pease, C.M. (1985) Biases in the durations and diversities of fossil taxa. *Paleobiology*, **11**, 272–292.

Pease, C.M. (1988a) Biases in total extinction rates of fossil taxa. *Journal of Theoretical Biology*, **130**, 1–7.

Pease, C.M. (1988b) Biases in per-taxon origination and extinction rates of fossil taxa. *Journal of Theoretical Biology*, **130**, 9–30.

Pease, C.M. (1988c) Biases in survivorship curves of fossil taxa. *Journal of Theoretical Biology*, **130**, 31–48.

Pease, C.M. (1988d) On comparing the geological durations of easily versus poorly fossilized taxa. *Journal of Theoretical Biology*, **133**, 255–257.

Pease, C.M. (1992) On the declining extinction and origination rates of fossil taxa. *Paleobiology*, **18**, 89–92.

Peterson, C.H. (1976) Relative abundances of living and dead molluscs in two Californian lagoons. *Lethaia*, **9**, 137–148.

Platnick, N.I. (1979) Philosophy and the transformation of cladistics. *Systematic Zoology*, **28**, 537–546.

Platnick, N.I. (1989) Cladistics and phylogenetic analysis today. In: B. Fernholm, K. Bremer & H. Jornvall (eds) *The Hierarchy of Life*, pp. 17–24. Excerpta Medica, Amsterdam.

Platnick, N.I., Griswold, C.E. & Coddington, J.A. (1991) On missing entries in cladistic analysis. *Cladistics*, **7**, 337–343.

Plotnick, R.E. (1986) Taphonomy of a modern shrimp: implications for the arthropod fossil record. *Palaios*, **1**, 286–293.

Pogue, M.G. & Mickevich, M.F. (1990) Character definitions and character state delineation: the bête noire of phylogenetic inference. *Cladistics*, **6**, 319–362.

Popov, L.E., Bassett, M.G., Holmer, L.E. & Laurie, J. (1993) Phylogenetic analysis of higher taxa of Brachiopoda. *Lethaia*, **26**, 1–5.

Ramskold, L. & Werdelin, L. (1991) The phylogeny and evolution of some phacopid trilobites. *Cladistics*, **7**, 29–74.

Raup, D.M. (1976a) Species diversity in the Phanerozoic: a tabulation. *Paleobiology*, **2**, 279–288.

Raup, D.M. (1976b) Species diversity in the Phanerozoic: an interpretation. *Paleobiology*, **2**, 289–297.

Raup, D.M. (1978) Cohort analysis of generic survivorship. *Paleobiology*, **4**, 1–15.

Raup, D.M. (1983) On the early origins of major biological groups. *Paleobiology*, **9**, 107–115.

Raup, D.M. (1986) Biological extinction in earth history. *Science*, **231**, 1528–1533.

Raup, D.M. (1987a) Mass extinction: a commentary. *Palaeontology*, **30**, 1–13.

Raup, D.M. (1987b) Major features of the fossil record and their implications for evolutionary rate studies. In: K.S.W. Campbell & M.F. Day (eds) *Rates of Evolution*, pp. 1–14. Allen & Unwin, London.

Raup, D.M. (1991) The future of analytical paleobiology. In: N.L. Gilinsky & P.W. Signor (eds) Analytical Paleobiology. *Short Courses in Paleontology*, **4**, 207–216.

Raup, D.M. & Boyajian, G.E. (1988) Patterns of generic extinction in the fossil record. *Paleobiology*, **14**, 109–125.

Raup, D.M. & Sepkoski, J.J. (1982) Mass extinctions in the marine fossil record. *Science*, **219**, 1239–1240.

Raup, D.M. & Sepkoski, J.J. (1984) Periodicity of extinctions in the geological past. *Proceedings of the National Academy of Sciences USA*, **81**, 801–805.

Raup, D.M. & Sepkoski, J.J. (1986) Periodic extinction of families and genera. *Science*, **231**, 833–836.

Raup, D.M., Gould, S.J., Schopf, T.J.M. & Simberloff, D.J. (1973) Stochastic models of phylogeny and the evolution of diversity. *Journal of Geology*, **81**, 525–542.

Rieppel, O. (1988) *Fundamentals of Comparative Biology*. Birkhäuser Verlag, Basel.

Rohlf, F.J. & Marcus, L.F. (1993) A revolution in morphometrics. *Trends in Ecology and Evolution*, **8**, 129–132.

Rose, K.D. & Bown, T.M. (1986) Gradual evolution and species discrimination in the fossil record. In: K.M. Flanagan & J.A. Lillegraven (eds) Vertebrates, phylogeny and philosophy. *Contributions to Geology, University of Wyoming Special Paper*, **3**, 119–130.

Rosen, D.E. (1979) Fishes from the uplands and intermontane basins of Guatemala: revisionary studies and comparative geography. *Bulletins of the American Museum of Natural History*, **162**, 267–376.

Roth, V.L. (1989) Fabrication noise in elephant dentitions. *Paleobiology*, **15**, 165–179.

Rowe, T. (1988) Definition, diagnosis, and origin of Mammalia. *Journal of Vertebrate Paleontology*, **8**, 67–83.

Russell, M.P. & Lindberg, D.R. (1988) Real and random patterns associated with molluscan spatial and temporal distributions. *Paleobiology*, **14**, 322–330.

Sadler, P.M. (1981) Sediment accumulation rates and completeness of stratigraphic sections. *Journal of Geology*, **89**, 569–584.

Sadler, P.M. & Strauss, D. (1990) Estimation of completeness of stratigraphical sections using empirical data and theoretical models. *Journal of the Geological Society, London*, **147**, 471–485.

Salthe, S.N. (1975) Problems of macroevolution (molecular evolution, phenotype definition and canalization) as seen from a hierarchical viewpoint. *American Zoologist*, **15**, 295–314.

Salthe, S.N. (1979) A comment on the use of the term 'emergent properties'. *American Naturalist*, **113**, 145–148.

Sanderson, M.J. (1990) Estimating rates of speciation and evolution: a bias due to homoplasy. *Cladistics*, **6**, 387–392.

Sanderson, M.J. & Bharathan, G. (1993) Does cladistic information affect inferences about branching rates? *Systematic Biology*, **42**, 1–17.

Sanderson, M.J. & Donoghue, M.J. (1989) Patterns of variation in levels of homoplasy. *Evolution*, **43**, 1781–1795.

Schaeffer, B. (1952) Rates of evolution in the coelacanth and dipnoan fishes. *Evolution*, **6**, 101–111.

Schindel, D.E. (1980) Microstratigraphical sampling and the limits of paleontologic resolution. *Paleobiology*, **6**, 408–426.

Schindel, D.E. (1982) Resolution analysis: a new approach to the gaps in the fossil record. *Paleobiology*, **8**, 340–353.

Schoch, R.M. (1986) *Phylogeny Reconstruction in Paleontology*. Van Nostrand Reinhold, New York.

Schopf, T.J.M., Raup, D.M., Gould, S.J. & Simerloff, D.S. (1975) Genomic versus morphologic rates of evolution: influence of morphologic complexity. *Paleobiology*, **1**, 63–70.

Sepkoski, J.J. (1978) A kinetic model of Phanerozoic taxonomic diversity, I. Analysis of marine orders. *Paleobiology*, **4**, 223–251.

Sepkoski, J.J. (1982) A compendium of fossil marine families. *Milwaukee Public Museum Contributions to Biology and Geology*, **51**, 1–125.

Sepkoski, J.J. (1984) A kinetic model of Phanerozoic taxonomic diversity, III. Post-Paleozoic families and mass extinctions. *Paleobiology*, **10**, 246–267.

Sepkoski, J.J. (1986) Global bioevents and the question of periodicity. In: O. Walliser (ed.) *Global Bio-events*, Lecture notes in earth sciences, Vol. 8, pp. 47–61. Springer-Verlag, Berlin.

Sepkoski, J.J. (1987) *Nature*, **330**, 252. (Reply to Patterson & Smith.)

Sepkoski, J.J. (1993) Ten years in the library: new data confirm paleontological patterns. *Paleobiology*, **19**, 43–51.

Sepkoski, J.J. & Raup, D.M. (1986) Periodicity in marine extinction events. In: D.K. Elliott (ed.) *Dynamics of Extinction*, pp. 3–36. Wiley, New York.

Sepkoski, J.J., Bambach, R.K., Raup, D.M. & Valentine, J.W. (1981) Phanerozoic marine diversity and the fossil record. *Nature*, **293**, 435–437.

Shao, K. & Sokal, R.R. (1990) Tree balance. *Systematic Zoology*, **39**, 266–276.

Shaw, A.B. (1964) *Time in Stratigraphy*. McGraw-Hill, New York.

Sheehan, P.M. (1977) Species diversity in the Phanerozoic: a reflection of labor by systematists? *Paleobiology*, **3**, 325–328.

Sheldon, P.R. (1987) Parallel gradualistic evolution of Ordovician trilobites. *Nature*, **330**, 561–563.

Signor, P.W. (1978) Species richness in the Phanerozoic: an investigation of sampling effects. *Paleobiology*, **4**, 394–406.

Signor, P.W. (1982) Species richness in the Phanerozoic: compensating for sampling bias. *Geology*, **10**, 625–628.

Signor, P.W. (1990) The geological history of diversity. *Annual Reviews in Ecology and Systematics*, **21**, 509–539.

Signor, P.W. & Lipps, J.H. (1982) Sampling bias, gradual extinction patterns and catastrophes in the fossil record. In: L.T. Silver & P.H. Schultz (eds) Geological implications of impacts of large asteroids and comets on the earth. *Geological Society of America Special Papers*, **190**, 291–296.

Simpson, G.G. (1944) *Tempo and Mode in Evolution*. Columbia University Press, New York.

Simpson, G.G. (1953) *The Major Features of Evolution*. Columbia University Press, New York.

Simpson, G.G. (1961) *Principles of Animal Taxonomy*. Columbia University Press, New York.

Skelton, P.W. (1978) The evolution of functional design in rudists (Hippuritacea) and its taxonomic implications. *Philosophical Transactions of the Royal Society, London*, **B284**, 305–318.

Skelton, P.W. (1985) Preadaptation and evolutionary innovations in rudist bivalves. In: J.C.W. Cope & P.W. Skelton (eds) Evolutionary case histories from the fossil record. *Special Papers in Palaeontology*, **33**, 159–173.

Skelton, P.W. (1991) Morphogenetic versus environmental cues for adaptive radiations. In: N. Schmidt-Kittler & K. Vogel (eds) *Constructional Morphology and Evolution*, pp. 375–388. Springer-Verlag, Berlin.

Skelton, P.W. (ed.) (1993) *Evolution, a Biological and Palaeontological Approach*. Addison Wesley, Wokingham.

Skelton, P.W., Crame, J.A., Morris, N.J. & Harper, E.M. (1990) Adaptive divergence and taxonomic radiation in post-Palaeozoic bivalves. In: P.D. Taylor & G.P. Larwood (eds) *Major Evolutionary Radiations*, pp. 91–117. Systematics Association Special Volume, No. 42. Clarendon Press, Oxford.

Smith, A.B. (1981) Implications of lantern morphology for the phylogeny of post-Palaeozoic echinoids. *Palaeontology*, **24**, 779–801.

Smith, A.B. (1984a) *Echinoid Palaeobiology*. George Allen & Unwin, London.

Smith, A.B. (1984b) Classification of the Echinodermata. *Palaeontology*, **27**, 431–459.

Smith, A.B. (1988a) Patterns of diversification and extinction in early Palaeozoic echinoderms. *Palaeontology*, **31**, 799–828.

Smith, A.B. (1988b) To group or not to group: the taxonomic position of *Xyloplax*. In: R.D. Burke, P.V. Mladenov, P. Lambert & R.L. Parsley (eds)

Echinoderm Biology, pp. 17–23. Balkema, Rotterdam.

Smith, A.B. (1989) Echinoid evolution from the Triassic to lower Jurassic. *Cahiers de l'Université Catholique de Lyon*, série science **3**, 79–117.

Smith, A.B. & Bengtson, P. (1992) Cretaceous echinoids from northeastern Brazil. *Fossils and Strata*, **31**, 1–88.

Smith, A.B. & Patterson, C. (1988) The influence of taxonomy on the perception of patterns of evolution. *Evolutionary Biology*, **23**, 127–216.

Smith, A.B. & Wright, C.W. (1990) British Cretaceous echinoids. Part 2, Echinothurioida, Diadematoida and Stirodonta (1, Calycina). *Palaeontographical Society Monographs*, **143** (583), 101–198, plates 33–72.

Smith, A.B. & Wright, C.W. (1993) British Cretaceous echinoids. Part 3, Stirodonta 2, Hemicidaroida and Phymosomatoida, part 1. *Palaeontographical Society Monographs*, **147** (593), 199–267, plates 73–92.

Smith, A.B., Lafay, B. & Christen, R. (1992) Comparative variation of morphological and molecular evolution through geologic time: 28S ribosomal RNA versus morphology in echinoids. *Philosophical Transactions of the Royal Society, London*, **B338**, 365–382.

Sober, E. (1983) Parsimony in systematics: philosophical issues. *Annual Reviews in Ecology and Systematics*, **14**, 335–357.

Sober, E. (1988) *Reconstructing the Past: Parsimony, Evolution and Inference*. M.I.T. Press, Massachusetts.

Sokal, R.R. (1986) Phenetic taxonomy: theory and methods. *Annual Reviews in Ecology and Systematics*, **17**, 423–442.

Sokal, R.R. & Crovello, T.J. (1970) The biological species concept: a critical evaluation. *American Naturalist*, **104**, 127–153.

Speyer, S.E. (1991) Trilobite taphonomy: a basis for comparative studies of arthropod preservation, functional anatomy and behaviour. In: S.K. Donovan (ed.) *The Processes of Fossilization*, pp. 194–219. Belhaven Press, London.

Springer, M.S. (1990) The effect of random range truncations on patterns of evolution in the fossil record. *Paleobiology*, **16**, 512–520.

Springer, M.S. & Lilje, A. (1988) Biostratigraphy and gap analysis: the expected sequence of biostratigraphic events. *Journal of Geology*, **96**, 228–236.

Sprinkle, J. (1983) Patterns and problems in echinoderm evolution. *Echinoderm Studies*, **1**, 1–18.

Sprinkle, J. & Moore, R.C. (1978) Echmatocrinea. In: R.C. Moore & C. Teichert (eds) *Treatise on Invertebrate Paleontology, Part T: Echinodermata 2*, pp. T405–407. Geological Society of America and University of Kansas Press, Lawrence, Kansas.

Stanley, S.M. (1979) *Macroevolution: pattern and process*. W.H. Freeman, San Francisco.

Stanley, S.M. & Yang, X. (1987) Approximate evolutionary stasis for bivalve morphology over millions of years: a multivariate, multilineage study. *Paleobiology*, **13**, 113–139.

Stanley, S.M., Signor, P.M., Lidgard, S. & Karr, A.F. (1981) Natural clades differ from 'random' clades: simulations and analysis. *Paleobiology*, **7**, 115–127.

Strathmann, R.R. (1991) Divergence and persistence of highly ranked taxa. In: A.M. Simonetta & S. Conway Morris (eds) *The Early Evolution of Metazoa and the Significance of Problematic Taxa*, pp. 15–18. Cambridge University Press, Cambridge.

Strauss, D. & Sadler, P.M. (1987) *Confidence Intervals for the Ends of Local*

Taxon Ranges. Technical report 158, Department of Statistics, University of California, Riverside, California. (Cited in Marshall, 1990.)

Strauss, D. & Sadler, P.M. (1989) Classical confidence intervals and Bayesian probability estimates for ends of local taxon ranges. *Mathematical Geology*, **21**, 411–427.

Surlyk, F. (1982) Brachiopods from the Campanian–Maastrichtian boundary sequence, Kronsmoor (NW Germany). *Geologisches Jahrbuch*, **A61**, 259–277.

Swofford, D.L. (1985) *PAUP, Phylogenetic Analysis Using Parsimony, Version 2.4.* (Computer program – IBM-PC compatible.)

Swofford, D.L. (1991) When are phylogeny estimates from molecular and morphological data incongruent? In: M.M. Miyamoto & J. Cracraft (eds) *Phylogenetic Analysis of DNA Sequences*, pp. 295–333. Oxford University Press, Oxford.

Swofford, D.L. (1993) *PAUP, Phylogenetic Analysis Using Parsimony, Version 3.1* (Computer program – Apple Macintosh compatible.)

Swofford, D.L. & Olsen, G.J. (1990) Phylogenetic reconstruction. In: D.M. Hillis & C. Moritz (eds) *Molecular Systematics*, pp. 411–501. Sinauer Associates, Sunderland, Massachusetts.

Szalay, F.S. (1977) Ancestors, descendants, sister groups and testing of phylogenetic hypotheses. *Systematic Zoology*, **26**, 12–18.

Tabachnick, R.E. & Bookstein, F.L. (1990) The structure of individual variation in Miocene *Globorotalia*. *Evolution*, **44**, 416–434.

Tassy, P. (1991) Phylogénie et classification des Proboscidea (Mammalia): historique et actualité. *Annales de Paléontologie (Vertébrés – Invertébrés)*, **76**, 159–224.

Taylor, P.D. & Larwood, G.P. (1990) Major evolutionary radiations in the Bryozoa. In: P.D. Taylor & G.P. Larwood (eds) *Major Evolutionary Radiations*, pp. 209–233. Systematics Association Special Volume, No. 42. Clarendon Press, Oxford.

Tegelaar. E.W., Kerp, H., Visscher, H., Schenck, P. & de Leeuw, J.W. (1991) Bias of the paleobotanical record as a consequence of variations in the chemical composition of higher vascular plant cuticles. *Paleobiology*, **17**, 133–144.

Temple, J.T. (1980) A numerical taxonomic study of the species of Trinucleidae (Trilobita) from the British Isles. *Transactions of the Royal Society of Edinburgh: Earth Sciences*, **71**, 213–233.

Temple, J.T. (1987) Early Llandovery brachiopods of Wales. *Monographs of the Palaeontographical Society*, **139** (572).

Temple, J.T. (1992) The progress of quantitative methods in palaeontology. *Palaeontology*, **35**, 475–484.

Templeton, A.R. (1989) The meaning of species and speciation. In: D. Otte & J.A. Endler (eds) *Speciation and its Consequences*, pp. 3–27. Sinauer Associates, Sunderland, Massachusetts.

Thackeray, J.F. (1990) Rates of extinction in marine invertebrates: further comparison between background and mass extinctions. *Paleobiology*, **16**, 22–24.

Tiffney, B.H. (1981) Diversity and major events in the evolution of land plants. In: K.V. Niklas (ed.) *Paleobotany, Paleoecology, and Evolution*, pp. 193–230. Praeger, New York.

Tipper, J.C. (1979) Rarefaction and rarefiction: the use and abuse of a method in paleoecology, *Paleobiology*, **5**, 423–434.

Tucker, M.E. (1991) The diagenesis of fossils. In: S.K. Donovan (ed.) *The Processes of Fossilization*, pp. 84–104. Belhaven Press, London.

Turner, A. & Paterson, H. (1991) Species and speciation: evolutionary tempo and mode in the fossil record reconsidered. *Geobios*, **24**, 761–769.

Ubaghs, G. (1967) Stylophora. In: R.C. Moore (ed.) *Treatise on Invertebrate Paleontology, Echinodermata 1(2)*, pp. S495–565. Geological Society of America and University of Kansas Press, Boulder, Kansas.

Valentine, J.W. (1969) Patterns of taxonomic and ecological structure of the shelf benthos during Phanerozoic times. *Palaeontology*, **12**, 684–709.

Valentine, J.W. (1989) How good was the fossil record? Clues from the Californian Pleistocene. *Paleobiology*, **15**, 83–94.

Valentine, J.W. (1990) The macroevolution of clade shape. In: R.M. Ross & W.D. Allmon (eds) *Causes of Evolution: a Paleontological Perspective*, pp. 128–150. University of Chicago Press, Chicago.

Valentine, J.W. & Jablonski, D. (1991) Biotic effects of sea level change: the Pleistocene test. *Journal of Geophysical Research*, **96**, 6873–6878.

Van Valen, L. (1973a) A new evolutionary law. *Evolutionary Theory*, **1**, 1–30.

Van Valen, L. (1973b) Are categories in different phyla comparable? *Taxon*, **22**, 333–373.

Van Valen, L. (1976) Ecological species, multispecies, and oaks. *Taxon*, **25**, 233–239.

Van Valen, L. (1978) Why not to be a cladist. *Evolutionary Theory*, **3**, 285–299.

Van Valen, L. (1984) A resetting of Phanerozoic community evolution. *Nature*, **307**, 50–52.

Van Valen, L. (1985) A theory of origination and extinction. *Evolutionary Theory*, **7**, 133–142.

Vrba, E.S. (1980) Evolution, species and fossils: how does life evolve? *South African Journal of Science*, **76**, 61–84.

Vrba, E.S. (1984) Evolutionary pattern and process in the sister-group Alcelaphini-Aepycerotini (Mammalia: Bovidae). In: N. Eldredge & S. Stanley (eds) *Living Fossils*, pp. 62–79. Springer-Verlag, New York.

Vrba, E.S. & Eldredge, N. (1984) Individuals, hierarchies and processes: towards a more complete evolutionary theory. *Paleobiology*, **10**, 146–171.

Vrba, E.S. & Gould, S.J. (1986) The hierarchical expansion of sorting and selection: sorting and selection cannot be equated. *Paleobiology*, **12**, 217–228.

Watrous, L.E. & Wheeler, Q.D. (1981) The out-group method of character analysis. *Systematic Zoology*, **30**, 1–11.

Westermann, G.E.G. (ed.) (1969) *Sexual Dimorphism in Fossil Metazoa and Taxonomic Implications*. Schweizerbart'sche, Stuttgart.

Westoll, T.S. (1949) On the evolution of the Dipnoi. In: G.L. Jepsen, E. Mayr & G.G. Simpson (eds) *Genetics, Paleontology and Evolution*, pp. 121–184. Princeton University Press, Princeton.

Westrop, S.R. (1991) Intercontinental variation in mass extinction patterns: influence of biogeographic structure. *Paleobiology*, **17**, 363–368.

Westrop, S.R. & Ludvigsen, R. (1987) Biogeographic control of trilobite mass extinction at an Upper Cambrian 'biomere' boundary. *Paleobiology*, **13**, 84–99.

Wheeler, Q.D. (1990) Ontogeny and character phylogeny. *Cladistics*, **6**, 225–268.

Wheeler, Q.D. & Nixon, K.C. (1990) Another way of looking at the species problem: a reply to de Queiroz & Donoghue. *Cladistics*, **6**, 77–81.

Wiley, E.O. (1978) The evolutionary species concept reconsidered. *Systematic Zoology*, **27**, 17–26.

Wiley, E.O. (1979) An annotated Linnaean hierarchy, with comments on natural taxa and competing systems. *Systematic Zoology*, **28**, 308–337.

Wiley, E.O. (1981) *Phylogenetics: The Theory and Practice of Phylogenetic Systematics*. Wiley, New York.

Willis, J.C. (1922) *Age and Area*. Cambridge University Press, Cambridge.

Willis, J.C. (1940) *The Course of Evolution*. Cambridge University Press, Cambridge.

Wilson, M.V.H. (1992) Importance for phylogeny of single and multiple stem-group fossil species with examples from freshwater fishes. *Systematic Biology*, **41**, 462–470.

Index

Page numbers in *italics* refer to figures and those in **bold** to tables.